Apollo in Perspective
Spaceflight Then and Now
Second Edition

T0138833

Apollo in Perspective
Spaceflight Then and Now
Second Edition

Jonathan Allday

CRC Press
Taylor & Francis Group
Boca Raton London New York

CRC Press is an imprint of the
Taylor & Francis Group, an **informa** business

CRC Press
Taylor & Francis Group
6000 Broken Sound Parkway NW, Suite 300
Boca Raton, FL 33487-2742

Printed on acid-free paper

International Standard Book Number-13: 978-0-367-26333-1 (Paperback)
978-0-367-26335-5 (Hardback)

Library of Congress Control Number: 2019943465

Visit the Taylor & Francis Web site at
http://www.taylorandfrancis.com

and the CRC Press Web site at
http://www.crcpress.com

*Ask ten different scientists about the environment, population control, genetics…
and you'll get ten different answers. But there's one thing every scientist on the
planet agrees on. Whether it happens in a hundred years, a thousand years
or a million years eventually our Sun will go cold and go out. When that happens,
it won't just take us – it'll take Marilyn Monroe, Lau Tzu, Einstein and
Buddy Holly and Aristophanes –all of this,all of this was for nothing.
Unless we go to the stars.*

Commander Jeffrey Sinclair
(from the science fiction series Babylon 5 by J M Straczinsky)

*You want to wake up in the morning and think the future is going
to be great – and that's what being a spacefaring civilization is all about.
It's about believing in the future and thinking that the future will be better
than the past. And I can't think of anything more exciting than going out
there and being among the stars.*

Elon Musk
SpaceX

*We're in a risky business … and we hope if anything happens to us, it will not
delay the program. The conquest of space is worth the risk of life. … Our God-given
curiosity will force us to go there ourselves because in the final analysis only man can
fully evaluate the moon in terms understandable to other men.*

Lieutenant Colonel Gus Grissom
*United States Air Force and
NASA Mercury 4, Gemini 3, and Apollo 1 Missions*

Contents

Foreword

I am very glad Jonathan Allday has undertaken this timely revision of *Apollo in Perspective*. The year 2019 represents the 50th anniversary of the first moon landing. I was 12 years old then, and as an aspiring astronaut I almost tried to jump into the television (so my mother told me later) as those first images of Neil Armstrong were sent around the world. That event and subsequent flights of Apollo spacecraft to the moon and Skylab reinforced my burning desire to fly in space – and travel to the Moon and Mars.

Fifty years later, I have flown in space six times, on the US Space Shuttle and on the Russian Soyuz, to the *Mir* Space Station and to the International Space Station, but never to the Moon or Mars. So as powerful as my dreams and desire were at the time, they were not fully realized. Instead other challenges and opportunities were presented to me, which is just fine. I am sure there is a woman or a man alive today who will land on the Moon and on Mars.

This book will certainly help them be ready for such a journey. Most importantly, it explains not only what happened 50 years ago, but how the Apollo missions happened, and the science that is required to do it again, or to go further, to Mars. If the reader is younger, still in school and perhaps considering the sciences, this book will introduce ideas that will help you choose the subjects to study that can help you to make your space travel a reality. For other readers, the book will be an exciting and thought-provoking read that gives a vision of the near future in space, which all of us on planet Earth will be able to enjoy as the adventure unfolds.

Jonathan Allday is a physics teacher who explains Newton's laws of motion, rockets, propulsion, rocket equations and orbits, all in a straightforward way. He fits those academic subjects into the adventure of space flight first aiming at Earth orbit, then *cis*-Lunar space, and on to Mars. He explains the importance of the Starliner and Dragon commercial crew vehicles, NASA's Orion and its new big booster space launch system, and how the goals of space faring nations have adapted to the 21st century. Now the emphasis on creating space programs and space infrastructure is to enable commercial activity to follow, letting international partnerships of many nations focus their limited budgets on the hardest exploration missions on the Moon and around Mars.

I enjoyed reading *Apollo in Perspective* as the story wove the exciting history of the Apollo program together with the physics and mathematics of space flight. I hope other readers value the book just as much.

Michael Foale, CBE
Former NASA Astronaut

Acknowledgments

Many thanks to Dr Kirsten Barr and Rebecca Davies at CRC Press/Taylor & Francis for patiently guiding me through two books in the same year. Also, thanks to the production team who turn this from a set of electronic documents into an actual book via their dark magic.

More than normal thanks go to my wife, Carolyn, who took over even more (all) of the domestic chores and managed me through the double duty of proofing one book while writing another.

I am very grateful to ex-NASA astronaut Dr Mike Foale for agreeing to write an introduction and also for the opportunities he gave to the staff and students at The King's School, Canterbury while he was onboard the ISS during Expedition 8 and also when he and his family visited the school during my tenure there. His visit also gave me the opportunity to meet the legendary TV commentator, Reginald Turnill, who was for many the voice of Apollo in the UK.

Finally, to the other writers, documentarists, film makers and enthusiasts who have kept the Apollo flame flying over the years, my thanks and admiration for the work that has been done.

Jonathan Allday
jallday40r@me.com
Canterbury, February 1999 (First Edition)
Bradford, July 2019 (Second Edition)

To the Moon by washing machine

Space has been part of my life ever since I was old enough to start reading the Captain W E Johns' books[1] (not *Biggles*, the science fiction ones chronicling the interplanetary adventures of Professor Lucius Brane and his friends). I had read Arthur C Clarke's *2001* (written with Stanley Kubrick) and had been totally baffled by the film (I was only 8…), so the late 1960s were a captivating period for me.

Watching the ghostly TV images from the Moon in the early morning of July 20, 1969, I did not feel any sense of history; I felt that space science and science fiction were coming true in front of my eyes. I can't imagine how anyone with a scientific bent and a lively imagination would not have been inspired.

The first edition of this book arose from a chance conversation with a sixth form (post 16) student at The King's School, Canterbury where I was working at the time. Having just watched a TV documentary about the faking of the Moon landings, this young person was adamant that the Apollo program had been an elaborate hoax. One reason he quoted revolved around the technology of the time. As the student pointed out, modern computers (that have advanced even further since the late 1990s) are much more powerful than those controlling the Apollo spaceflight. In fact, the processors running a modern washing machine would run rings around the Apollo systems. While discussing the washing machine example, my mind flitted to a scene from the movie *Apollo 13* in which Jim Lovell's mother makes a stirring remark: 'if they could make a washing machine fly, my Jimmy could land it'. Hence I determined to write a book called 'To the Moon by Washing Machine' to try and explain how the technology of the time was sufficient to the task, demonstrate some of the basic physics required and remind people that the Apollo astronauts were pilots: very good ones.

The book turned out to be about more than just the Apollo program and acquired a different name. It was about Apollo, judged from the perspective at the end of the 20th century. From this vantage point, progress seems halting and hesitant since the heady days of the 1960s, when scientists seriously thought that they could have a man on Mars by the 1980s.

Now it is 50 years since Apollo and some 20 years since the first edition (how did that happen?). Much has transpired since then. The *Columbia* tragedy added to the toll of deaths from Apollo 1 and continued through the *Challenger* catastrophe.[2] The Space Shuttle has been retired and the International Space Station (ISS) all but completely assembled.

I had the privilege of being involved with a live radio link between the students of The King's School and the ISS when alumni astronaut Mike Foale commanded Expedition 8 (October 2003 through April 2004). On February 6, 2018 I once again found myself cheering a rocket, as SpaceX's *Falcon Heavy* launched on its maiden flight.

From time to time, my mind has turned to writing again about space, so when updating this book for the 50th anniversary of Apollo 11 was suggested, it was a chance I could not miss.

In this second edition I have refined some of the content, used a wider range of images, corrected some mistakes and some misinformation that existed at the time of the first edition, and also looked back on Apollo with the further perspective of 20 years and America's newly declared intention of returning to the Moon. In this context, commercial space flight and the ambitions of a few wealthy individuals (principally, but not only, Elon Musk) added a new angle that I don't think was foreseen, even at the time of the first edition.

I have added a chapter on the achievements of the Apollo missions and removed a chapter from the first edition that dealt with the prospects of travelling to the stars. I hold by the content of that chapter, although it seems out of place in the current modern context. An undated version will appear in due course on the Taylor & Francis web site.

One thing certainly remains true from the time of the first edition. Now that little seems to be beyond us, technologically speaking, it is important to try and maintain the exciting sense of pushing technology to the edge. How else will we motivate students to consider engineering and the technological sciences, let alone inspire them to become space explorers? For me, this is the abiding unfortunate consequence of the decision of the United Kingdom (UK) not to take part in the European Space Agency's manned missions to the ISS. The Internet has made this a world with significantly fewer boundaries, but I can't help feeling that it is a shame that our young scientists and engineers have no UK astronauts as role models.

Perhaps I will get the chance to sit with my family and watch the first steps onto the surface of Mars, but I wonder whether the mission patch will signify a country or a company.

ENDNOTES

1. He died in 1968, so did not see Apollo 8 or the Moon landing.
2. Of course, other lives have been lost in space exploration as well. I focus here only on US missions.

Apollo in outline

1.1 THE POLITICS OF APOLLO

On May 25, 1961, John F Kennedy made a speech to the US Congress committing his government to land a man on the Moon before the end of the 1960s. The president was inspirational, full of passion, and promised that this 'greatest of all adventures' would rebuild the country's confidence, so badly dented by Soviet successes in space:

> No single space project in this period will be more exciting, or more impressive to mankind, or more important for the long-range exploration of space; and none will be so difficult or expensive to accomplish …

In taking over the presidency from Eisenhower, Kennedy inherited a manned programme (Mercury) that was struggling to compete with the Soviet achievements and a scientific community that was deeply divided over the value of manned space exploration.

At the time Kennedy delivered his congressional speech on 'Urgent National Needs', only Alan B Shepard's 15-minute flight had grazed the limits of space on May 5, 1961. This was about a month after a Soviet mission in which Yuri Gagarin made one complete orbit of the Earth (April 12, 1961). The first American into Earth orbit was John H Glenn, who circled the Earth three times on February 20, 1962.

Both Eisenhower's and Kennedy's chief scientific advisors felt that the results of manned flight could not compare with the likely benefits of an unmanned satellite programme. Indeed Eisenhower, in his departing budget speech (January 18, 1961), said that more work would be needed to determine whether any benefit would result from extending the manned space effort beyond the Mercury programme.

NASA engineers were confident that they could land a man on the Moon, given sufficient development time. Much scientific research could be done with unmanned probes, and satellite technology was set to revolutionise communications and weather forecasting, but for exploring space, NASA was convinced that people had to fly. In October 1960, NASA issued contracts to three aerospace firms to study the feasibility of a lunar mission. However, this research and planning would be in vain without the availability of substantial funding and that required presidential and congressional backing. Trying to justify such expenditure in terms of pure science was not going to work. Human exploration of the Moon, no matter how poetic and important to the human spirit, was not going to reap any greater scientific rewards than automatic probes. After all, the people being trained to fly were not scientists.

In the end, NASA's ambitions had to be realised on non-scientific grounds. The climate of competition with the Soviets was just what was needed to put pressure on the president.

Kennedy, as President-Elect, called on his Vice President-Elect, Lyndon B Johnson (who was an enthusiastic supporter of the space programme) to conduct an investigation into the nation's space efforts, and find a project that would produce 'dramatic results' and demonstrate US superiority in space. Johnson's report was presented on May 8, 1961. It recommended expanding the space programme, commissioning a new rocket to lift heavier payloads into space and committing NASA to land a man on the Moon. The result was Kennedy's speech to Congress. Space was not the only issue covered, but the president's strong statement will always be remembered as the moment that America decided to go to the Moon.

Congressional support was not assured, but members turned out to be solidly behind the idea. Kennedy asked to increase Eisenhower's $1.1 billion budget for the space programme by $675 million. He got virtually all of that money and Apollo was underway.

1.2 THE MISSION

The plan to land men on the Moon was audacious: a challenge to technology, skill and endurance.

A Saturn V rocket lifted a three-man crew and their equipment from the surface of Earth (Figure 1.1). This booster had three sections (called *stages*), with independent fuel tanks and rocket engines. Each stage burnt for a set time, after which explosive charges fired to separate the spent casing from the rest of the rocket. As a result, the stack became progressively lighter during the flight. Surprisingly, orchestrating a launch in this way uses less fuel than deploying a single stage, which more than compensates for the extra mass of the multiple engines and fuel tanks needed. We will discuss this aspect of rocketry in more detail in Section 3.6.

By the time the third stage's engine had shut down, the Apollo astronauts were in orbit. After some time circling the Earth, while the crew and ground support teams performed a thorough programme of checks, the third stage's engine fired again to inject the craft into a much larger orbit which intersected with that of the Moon. For the first few lunar missions (Apollos 8, 10 and 11), the path had been designed to swing the spacecraft around the far side of the Moon and return it to Earth. This *free return trajectory* was a fail-safe mechanism to bring the astronauts home automatically if their main engine failed to fire to put them in orbit around the Moon. Later missions deviated from free return, as it restricted the regions that were accessible for landing sites.

Soon after this *translunar injection* 'burn', the *command module* (in which the crew lived) attached to the *service module* (which contained fuel, air and electrical power generating equipment, as well as an engine) separated from the third stage of the booster. The command module pilot manoeuvred the combination

FIGURE 1.1 The launch of the Apollo 11 Saturn V booster. (Image courtesy of NASA; scan courtesy of Kip Teague.)

(known as the command and service module or CSM) out to a safe distance. Once clear, he turned it around and brought it back, head-first, towards the third stage, nestling inside which was the *lunar module* (LM[1]): the craft designed to land on the surface of the Moon. Throughout the launch, this fragile vehicle was covered by shielding at the top of the third stage and under the CSM (Figure 1.2). Now it was necessary to join the Apollo elements together by docking the command module with the lunar module and slowly extracting it from its housing (see Figure 5.13).

Finally, the joined craft backed away from the third stage: its direct part of the mission was over. The redundant third stages for the Apollo 8 through 12 missions were placed into an orbit around the Sun; while on the later missions (Apollo 13 through 17) the third stages were targeted to crash on the Moon in order to give lunar geologists valuable seismic information.

Some 3 days after launch, the Apollo CSM/LM combination arrived in the vicinity of the Moon. To enter lunar orbit, the spacecraft had to slow down by firing the service module engine as it passed behind the Moon (Figure 1.3). Consequently, one of the most critical manoeuvres in the mission was performed while the astronauts were unable to contact Earth. Mission control would not know whether the burn had been successful until a radio link was established again, as Apollo crept around the edge of the Moon. The precise time delay between the moment at which contact was lost and then re-established had been carefully calculated on the basis of a successful burn. As a result, all eyes were on the clock in mission control while the team waited to see whether the astronauts safely entered lunar orbit.

Once they were securely circling the Moon, the mission commander and lunar module pilot floated into the LM to power it up and check out its systems.[2] The command module pilot remained with the CSM, while his two colleagues detached the lunar module and prepared for the landing.

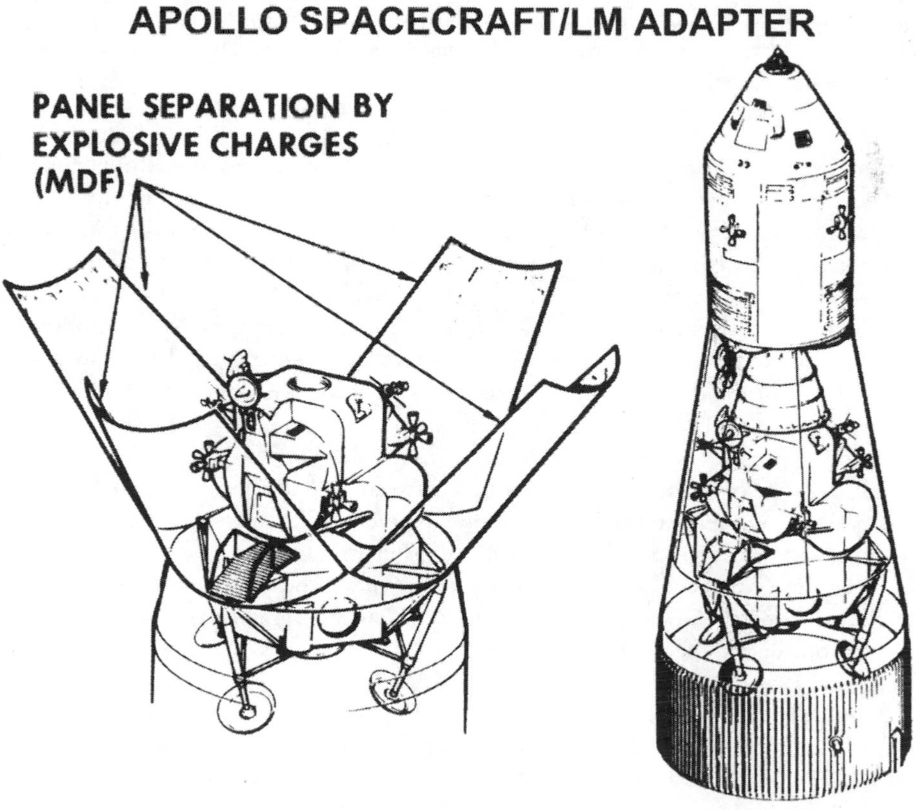

APOLLO SPACECRAFT/LM ADAPTER

PANEL SEPARATION BY EXPLOSIVE CHARGES (MDF)

FIGURE 1.2 The lunar module was housed beneath the CSM at the top of the Saturn V third stage during the launch. (Image courtesy of NASA.)

FIGURE 1.3 The Apollo 15 command and service module combination in lunar orbit. The command module is the blunt cone at the right end of the cylindrical service module. A panel in the SM has opened so that a camera can take pictures of the lunar surface. The large service module engine bell can be seen. This engine handled insertion and extraction of the craft from lunar orbit. The picture was taken from the lunar module during rendezvous after the landing. (Image courtesy of NASA.)

From his vantage point, the astronaut in the command module was able to give the lander a visual inspection while it manoeuvred in front of him (Figure 1.4). The LM's descent engine was then fired, taking it into a lower orbit while further checks and simulations were carried out. The LM crew spent a couple of orbits on such procedures, including getting used to the sequence of lunar features that would appear along the landing trajectory.

Finally, the descent engine was fired once again to put the LM into *powered descent*. Under engine power, the two astronauts descended to a pre-defined landing area. However, as the best available photographs of the Moon could not resolve all its features, the final choice of the least hazardous landing spot was left open to the commander flying the lunar module.

Neil Armstrong (commander) and Buzz Aldrin (lunar module pilot) carried out the first landing. From the start, there were communications difficulties between ground control and the lunar module. At times messages had to be relayed through the command module. Then, during the powered descent, other problems began to emerge. Although the trajectory from orbit to the surface had been calculated and programmed in advance, vital factors were unknown. For example, local regions of Moon rock with higher than average densities pulled on the LM with a greater than expected gravitational force, with potential to deflect the vehicle off course. There was no way to know about these mass concentrations (*mascons*) before flying over one.[3]

Armstrong noticed that various landmarks were appearing some 2 to 3 seconds before he expected them: the LM was travelling more quickly than anticipated. At the speed that they were moving, 2 or 3 seconds too early indicated that they would overshoot the landing spot by about 5 km. However, as the onboard computer was unaware of the issue, the programmed throttle-down came on time, rather than slightly late to compensate for the faster ground speed. Furthermore, the computerised path had the lunar module landing in a region full of boulders, some of which were as large as cars.

FIGURE 1.4 The Apollo 11 lunar module during inspection manoeuvres prior to landing. This picture was taken from the command module by its pilot (Mike Collins). (Image courtesy of NASA.)

Armstrong took manual control and started looking for a safer place to land. To add to the pressure, the LM computer was periodically displaying 'program alarms' – the equivalent of error messages that one gets (all too often) on today's PCs. Due to rogue signals in the rendezvous radar (used to locate the CSM for redocking), spurious data was being sent to the computer. Between running the landing and trying to deal with data coming from the radar systems, the computer was given too much to do and was 'overflowing'. Fortunately, the mission control experts were able to confirm that the alarms did not represent a serious problem and that the computer was simply signalling that it had too much work and would clear some jobs to focus on vital tasks needed for landing.[4] The software would restart safely after each alarm.

Armstrong continued to search for a suitable landing spot and the tension in mission control was climbing. The mission rules set an upper limit on the amount of fuel that could be used during the descent: if the LM had less than 20 seconds of fuel left, the standing order was to pull out of the landing. Armstrong would have to hit the abort button, triggering a process that would cut the descent engine and separate the LM's ascent stage from the descent stage. The ascent engine would fire, adding velocity to the descending craft in an attempt to achieve a stable orbit. However, as the astronauts indicated post-flight, the 'correct' reaction during a training exercise is not necessarily at the front of their minds during an actual mission when they are motivated to succeed. As long as they were low enough they could always fall the rest of the way to the surface, if they ran out of fuel.

As the descent engine finally shut down and the Apollo 11 lunar module settled onto the Moon, the crew thought they had only 25 seconds of fuel remaining, although subsequent analysis showed that this was nearer to 45 seconds. When Armstrong had to pitch the LM over to fly past the boulders, fuel sloshing around in the tanks had upset the readings.

In the moments after the landing, and while much of the world celebrated, both mission control and the men on the Moon were busy performing system checks to determine whether it was safe to remain[5].

According to the mission timeline, at 102:45:58, Armstrong made his famous announcement, *'Houston, Tranquility Base here. The Eagle has landed.'* Not until 102:56:02 did the world get the first description of the surface from Aldrin: *'It looks like a collection of just about every variety of shape, angularity, granularity, about every variety of rock you could find.'*

In total, Armstrong and Aldrin lived, worked and slept on the Moon for 21 hours and 36 minutes, 2.5 hours of which were spent on the surface within 90 m of the lunar module (Figures 1.5 and 1.6). The later missions expanded the exploration time significantly. The Apollo 17 crew spent 3 days on the Moon and ventured out of their craft for a total of 22 hours, driving a vehicle known as the *lunar rover* to cover 35 km. All the Apollo astronauts were trained in geology, but it was not until the final mission that Harrison Schmitt, a professional geologist trained as an astronaut, was sent to the Moon.

To start the return journey the LM's ascent engine fired, separating the top half containing the crew cabin from the bottom of the craft with the descent engine and the landing legs. The ascent engine carried them back into orbit and a rendezvous with the command module pilot orbiting in the CSM. After docking, the astronauts transferred their Moon rock samples, cameras and film back to the CSM and jettisoned the now redundant ascent stage of the LM.

Once again, behind the Moon, the CSM engine fired to break the craft out of orbit onto a trajectory that brought it back to Earth. The return journey took another 3 days.

Shortly before arriving, the astronauts jettisoned the service module and turned the command module around so that its underside faced the Earth. Strapped into couches, they entered the atmosphere backs first. Beneath them, the lower surface of the command module consisted of a specially designed shield that protected the capsule and its occupants from the heat of re-entry.

The craft hit the atmosphere at 40,000 km per hour. The temperature rapidly climbed to 5,000°C as friction transformed kinetic energy into heat. Contact was lost with the ground as the high temperatures turned the atmospheric layer surrounding the module into a highly charged plasma that was impenetrable to radio waves. The CM emerged from its entry immolation travelling at about 480 km per hour at an

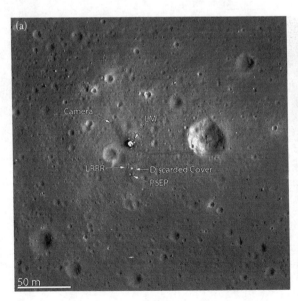
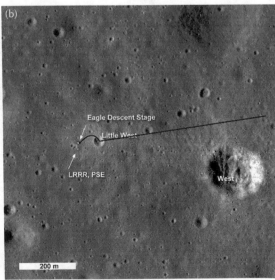

FIGURE 1.5 (a) The Apollo 11 landing site as imaged by the Lunar Reconnaissance Orbiter in March 2012. The lunar module descent stage is clearly visible. The tracks from Armstrong's stroll to the Little West Crater depicted in the film 'First Man' is the unlabelled straight feature to the right of the LM. Also visible are the Lunar Ranging Retro Reflector and Passive Seismic Experiment installation. (b) Wider field view of Apollo 11 landing site. The black line was added to image to indicate roughly the direction in which the lander traversed the lunar surface. Armstrong took manual control to avoid landing in the vicinity of the boulder field north of the crater. (Image courtesy of NASA, Goddard Space Flight Center, and Arizona State University.)

FIGURE 1.6 The Apollo 11 landing site showing the lunar module, the PSE in the foreground (next to the astronaut) and the Lunar Ranging Retro Reflector just behind it. (Image courtesy of NASA.)

altitude of 10.7 km. Three large parachutes were then deployed (Figure 1.7) and the capsule floated down into the sea. The astronauts were then collected by helicopter and flown to an aircraft carrier.

Although scientists were quite confident that no microorganisms could live on the Moon, it was deemed safest to keep the crew in quarantine for a period to see whether any diseases developed. Consequently, when they alighted from the helicopter onto the deck of an aircraft carrier, they wore full biological isolation suits and walked straight into a quarantine chamber where they lived for the next 2 weeks. This practice continued through to Apollo 15. Later missions ended with the astronauts climbing from helicopters in fatigues and baseball caps.

In all, there were eleven manned Apollo missions, nine of which went to the Moon. Apollo 8 took the command and service modules into orbit around the Moon. Apollo 10 carried a lunar module along as well and flew it down to 9 miles above the surface. Eight of the crews departed with the intention of landing. However, due to the accident onboard Apollo 13, only seven landings took place. The original plan was to continue the programme through to Apollo 20, but declining public interest and American involvement with other budget-draining activities (e.g. the Vietnam War) led to pressure on NASA and the missions ended with Apollo 17.

1.3 THE MOON

As the brightest object visible in the night sky, it is hardly surprising that the Moon has been a major feature in human imagination.[6] In 1865, Jules Verne wrote a story about explorers travelling to the Moon in a shell shot from a giant cannon (*From the Earth to the Moon*). Although the technology is impossible, the theme of lunar exploration was established early in the history of science fiction.

FIGURE 1.7 The Apollo 16 command module shortly before splashdown. (Image courtesy of NASA.)

The most conspicuous aspect of the Moon as we see it from Earth is the manner in which it waxes and wanes in a regular monthly cycle. As the Moon circles the Earth, one side is always lit by sunlight and the phases of the Moon result from our changing perspectives of its face.

The surface features of the Moon do not alter from day to day: we see more or less of the same disc as the Moon's crescent changes. This probably does not take most people by surprise, yet it is a curious fact. The Moon is slightly egg-shaped, with the pointed end directed towards the Earth. The action of the Earth's gravitational force on the Moon has, over the history of the solar system, equalized the time required for the Moon to turn on its own axis with the time taken for it to orbit the Earth. Without this phenomenon, the reverse side of the Moon would be presented to us as it turned on its axis.

The first humans to see the far side of the Moon with the naked eye were the three astronauts (Borman, Lovell and Anders) who took part in the Apollo 8 mission.[7] The similarity between their names and the ones chosen by Verne for his three heroes (Barbicane, Nicholl and Ardan) was an irony that was not lost on the Apollo 8 crew. Neither was the fact that Verne left his characters stranded in orbit around the Moon at the end of his novel.[8]

The near side of the Moon (Figure 1.8) is dominated by large dark areas called *maria* (seas; singular is *mare*) and the bright regions of the lunar highlands. Without the aid of telescopes, ancient lunar observers thought that the dark areas were seas of water, as on Earth. Consequently, they named them appropriately, for example, *Oceanus Procellarum* (Ocean of Storms), *Mare Serenitatis* (Sea of Serenity) and most famously *Mare Tranquillitatis* (Sea of Tranquility) where Apollo 11 landed. In total, 16% of the Moon's surface is covered by the maria, mostly on the near side. Some are very large. The Ocean of Storms has an area bigger than that of the Mediterranean.

FIGURE 1.8 The near side of the Moon as seen from Earth. This is the face that we always see. The dark maria are evident along with the brighter (and more cratered) lunar highlands. (Image courtesy of NASA, Goddard Space Flight Center, Arizona State University and Lunar Reconnaissance Orbiter.)

As the maria are relatively lightly cratered compared with the highly pockmarked highlands, we can deduce that they represent younger surfaces of the Moon. Over the millions of years since its formation, the Moon has been bombarded with meteorites. Unlike the Earth, the Moon has no atmosphere to protect its surface, so the crust is fragmented and covered by holes left by meteorite strikes (craters). Sometimes the material thrown from the surface can be seen as long streaks or *rays* spreading out from a crater (Figure 1.9).

About 4 billion years ago, several huge impacts occurred, forming vast craters. Then around 3.5 billion years ago, active volcanoes produced lava flows that filled the impact basins with a much darker coloured iron-rich basalt material, forming the maria. In the process, the lava covered old craters that had been formed. Since then, a few more meteorite impacts have marked the surfaces of the seas, but the highlands that escaped the lava floods show a complete history of bombardments dating back to the Moon's formation.

The rocks found in the highlands are at least 4.5 billion years old, so they formed at about the same time as the Earth. This makes them the oldest rocks on the Moon. They are rich in a light-coloured mineral called feldspar which gives the lunar highlands their bright colour.

Not all meteorites are large enough to create craters. Much smaller ones, called *micrometeorites*, have continually rained down on the Moon over the millennia. Their bombardment thoroughly pulverised the rocks on the surface, leaving rock fragments and a fine dust-like material that together are referred to as the *regolith*.

On Earth, regolith is formed from the *weathering* of rocks. Physical weathering happens when water in the rocks freezes and expands to cause cracks. Chemical weathering comes about when rock minerals

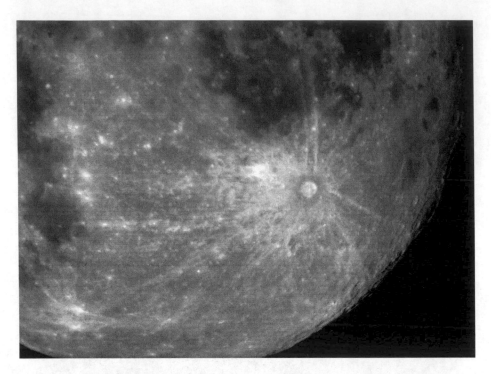

FIGURE 1.9 Rays spreading out from Tycho crater. (Image courtesy of Brian, https://www.flickr.com/people/13447091@N00, Bountiful, Utah, under CCA 2.0: https://creativecommons.org/licenses/by/2.0/deed.en.)

dissolve in water or acid. Biological weathering takes place when growing plant roots open up cracks in the rock. All of these are impossible on the Moon due to the lack of atmosphere, water and wind.

The lunar regolith covers all of the Moon's surface except for the steep walls of craters or valleys. The regolith is 2 to 8 m thick over the maria and possibly exceeds 15 m on the highlands, depending on how long the rock underneath it has been exposed to bombardment.

The lunar far side is much closer in character to the highlands of the near side (Figure 1.10). Although there are basins comparable in size to some of the major maria on the near side, they are not filled with the same dark material. Lunar geologists discovered that the centre of gravity of the Moon is not quite at its geometrical centre – it is about 2 km nearer to the Earth.[9] This may be because the Moon's *crust* is thicker on the far side. Crustal material is less dense than the underlying *mantle*. So, if there is a greater thickness of crust on the far side of the Moon's geometrical centre, there is also less mantle than on the near side; hence the near side is heavier. This ties in with the lack of maria on the far side, despite the presence of suitable basins to fill. A thicker crust would make it difficult for lava to reach the surface.

1.3.1 APOLLO'S CONTRIBUTION TO LUNAR SCIENCE

Before the Apollo landings, there was a great deal of speculation about the nature of the Moon, but little in the way of definite information. While they were on the surface, the Apollo astronauts carried out geological surveys and sampled rocks that were later brought back to Earth for analysis. As a result of their work, the Apollo programme was directly responsible for establishing some basic facts about the Moon:

- The Moon is made of rocky material. Various rocks display evidence that the lunar material has been melted, squirted from volcanoes, and crushed by meteorite impacts. The Moon has a thick crust (60 km), a fairly uniform *lithosphere* (60 to 1,000 km) and a partly liquid *asthenosphere* (1,000 to 1,740 km) corresponding to equivalent layers found within the Earth. The Moon may have a small iron core. Some rocks show that the moon had a magnetic field in the past, although no lunar field remains today.
- The Apollo astronauts brought back rock samples that enabled various craters to be dated. Other planets, such as Mercury, Venus and Mars, had periods of crater formation similar to those found on the Moon.

FIGURE 1.10 The lunar far side that is never seen from Earth. The character of most of this surface is much closer to that of the highlands on the near side. (Image courtesy of NASA, Goddard Space Flight Center, Arizona State University and Lunar Reconnaissance Orbiter.)

Now that we have dates for the craters on the Moon, we can draw conclusions about the ages of similar craters on these planets. Before Apollo, however, the origins of lunar impact craters were not fully understood and the causes of similar craters on Earth were highly debated.

- Moon rock ages range from about 3.2 billion years in the maria to nearly 4.6 billion years in the highlands. On Earth, plate tectonics and erosion continually recycle rocks. No such processes exist on the Moon, so ancient surfaces, are still visible. The oldest rocks found on Earth's surface are about the same age as the youngest rocks found on the Moon. Consequently, the Moon has a far longer record of the events that must have affected both bodies.

- The Earth and the Moon have a common ancestry. Studies of the amounts of various oxygen isotopes found in Moon rocks and in rocks on Earth show that the proportions are very similar. However, the Moon rocks are highly depleted in iron and the elements needed to form atmosphere or water.

- Extensive testing revealed no evidence of life, past or present, on the Moon. In fact, remarkably little evidence was found for organic compounds that can form without biological processes being at work. Such chemicals are regularly found in meteorites striking Earth, and the small amounts found on the Moon may be the results of meteorite strikes as well.[10]

- The three types of rock found on the Moon are *basalts*, *anorthosites* and *breccias*. Basalts are the dark lava flows that fill mare basins. They are similar to the lavas from which the Earth's ocean crusts are made but much older. Anorthosites are the light rocks from which the highlands are formed. They contain large amounts of feldspar. Breccias are composite rocks formed from all other rock types through crushing, mixing and smashing by meteorite impacts. Significantly, no sandstones, limestones or similar rocks were found on the Moon. Such rocks on Earth were formed from sediments deposited by flowing water.

- Early in its history, (4.4 to 4.6 billion years ago) the Moon was molten to great depths. As this 'magma ocean' slowly cooled, minerals started to crystallize within it. The earliest ones to form (those with the highest melting points) were comparatively dense and tended to sink as they solidified. The low melting point materials crystallized later near to the surface, being less dense, and produced the lunar highlands. Over billions of years, meteorite impacts reduced much of the ancient crust, leaving the mountain ranges that we now see between the basins.
- The large dark basins are gigantic impact craters filled by lava flows about 3.2 to 3.9 billion years ago.
- Large mass concentrations (mascons) lie beneath many lunar basins. These mark places where lava formed thick layers under the surface.
- The Moon's shape is slightly asymmetrical, probably due to the continual pull of Earth's gravity during its evolution. Its crust is thicker on the far side, while most volcanic basins and unusual mass concentrations occur on the near side.
- Radiation pouring from the Sun acts on the regolith and surface rocks of the Moon, changing their chemical compositions. Consequently, the Moon contains a 4 billion year record of the Sun's history. We are unlikely to find such a complete account on any other planet.

1.3.2 THE ORIGIN OF THE MOON

Four theories have been put forward to explain how the Moon was formed. Each of them has to deal with two significant facts. First, among all the planet and Moon combinations in the solar system, our Moon is by some margin the largest compared to the size of the planet it orbits. Second, the Moon contains a much smaller percentage of iron than the Earth.

Each theory relies on the well-accepted model of how the Sun and planets formed. About 4.5 billion years ago, a large cloud of gas slowly contracted under gravity into the glowing ball that became the Sun. As the atoms in the cloud were pulled together by their mutual gravitational attraction, they picked up speed (just as a falling rock will accelerate). Collisions between the atoms ensured that they moved about at random. In a gas, the average kinetic energy of molecules in random motion is what we measure as temperature. As the gas cloud collapsed, the collisions brought about an increase in temperature, which eventually reached a level sufficient to trigger nuclear fusion. The energy produced by these reactions served to increase the temperature further and the cloud stopped contracting as the atoms were moving fast enough to resist the inward pull of gravity (this is the natural tendency of any hot gas to expand). Currently, the Sun is enjoying comfortable middle age, settled in a balance between the gravitational force of its mass trying to collapse it down and the tendency of its hot gas to expand. This balance will continue for something like another 4.5 billion years.

As the central part of the gas cloud contracted to form the Sun, some of the heavier material left behind formed lumps of rock called *planetoids*. By colliding with and pulling on one another via gravitational forces, the planetoids eventually merged to form the rocky planets that now exist in the solar system. The surfaces of these planets were very hot and continually battered by meteorites, which prevented much in the way of cooling from taking place. This is how the Earth formed.

Turning to the Moon's origin, one theory suggests that the Earth was spinning very fast and threw off a large lump of material that became the Moon (rather like mashed potato flying off a plate that is spun too quickly[11]). Provided this happened after the iron in the cooling Earth had sunk towards its centre (as iron would, being heavier than the rest of the material), this could explain the lack of lunar iron. However, there is a considerable snag with this theory. If you calculate how fast the Earth's spin would be if it re-absorbed the Moon, you will find that the Earth would rotate on its axis once in 8 hours (rather than 24 hours). Unfortunately, this very rapid rotation rate is not enough to account for throwing off a lump of material the size of the Moon. That would require a spin rate sufficient to shorten a day to something like 2 hours. Consequently, unless some way can be found to explain how a great deal of energy was lost, this theory does not work.[12]

The second theory suggests that the Moon and the Earth arose at the same time from planetoids and rather than merging, they went into mutual orbit. This offers no explanation for the very different metallic structures of the two bodies.

Another theory argues for the Moon having formed somewhere else and being captured, as it passed, by the Earth's gravity. This hypothesis suffers from two snags. The first is that the Moon would have to lose energy in order to be snared by the Earth in this manner, and there is no obvious mechanism to account for this loss. The second is that aside from iron content, the Earth and the Moon are too similar to suggest that they come from totally different places.

The fourth theory is given the most credence. According to this view, at an early point in the Earth's formation, it collided with a very large object (probably about the size of the planet Mars). This massive impact would have vaporised the object and much of the Earth's crust. In this scenario, slow moving material falls back to Earth while quicker debris escapes into space. The rest would form a ring of hot gas around the Earth. Eventually, gravity would pull the ring material together to become the Moon. If the collision took place after the Earth's iron-rich core formed, the outer material blasted away would have been poor in iron content, as is the Moon today. It is possible that the collision also resulted in the 23.5° tilt of the Earth's axis, resulting in our seasons.

Despite its convincing aspects, the theory still has problems. In 2001, a detailed analysis of Moon rocks showed that their isotopic signature (the proportions of various isotopes found in samples) was identical with that of Earth and very different from those of other bodies in the solar system. This was seen as a surprising result, as most of the debris ring from which the Moon is thought to have formed should have come from the colliding object, not from the Earth. Further work in 2007 demonstrated that the chances of the colliding object having the same isotopic signature as that of Earth are too small to represent a serious consideration.

To account for these observations, a variant theory (2012) was proposed: two bodies five times the size of Mars collided and then re-collided to form a debris ring from which the Earth and the Moon coalesced in mutual orbit.[13]

Clearly, the origin of the Moon is more complex than we thought, and research is ongoing.

1.3.3 MOON FACTS

Mass (Earth = 1)	0.012
Radius at equator (Earth = 1)	0.27
Average distance from Earth (km)	384,400
Time to turn on axis (days)	27.32166
Time to orbit Earth (days)	27.32166
Gravitational acceleration at equator (m/s^2)	1.62
Average surface temperature (day)	107°C
Average surface temperature (night)	−153°C
Maximum surface temperature	123°C
Minimum surface temperature	−233°C

1.4 THE IMMEDIATE FUTURE

A detailed look at NASA's current plans for a return to the Moon will be the subject of Chapter 9. However, the recent discovery of water ice on the Moon confirmed decades of speculation and has influenced the choice of landing sites for further Moon missions.

In September 2009, it was announced that the Moon Mineralogy Mapper, an instrument on the *Chandrayaan 1* satellite launched in 2008 by the Indian Space Research Organisation, found evidence of water ice on the Moon.[14] In 2018, the results of further analysis of the data led to a confirmed discovery[15] of water ice at the lunar poles. This makes the polar region, and the south lunar pole especially (where the ice concentration is higher), excellent sites for establishing lunar bases.

In certain regions around the poles, sunlight is accessible at all times, which is useful for generating solar power and for long duration excursions. Furthermore, water ice in craters that remain permanently in shadow could provide oxygen and hydrogen (via electrolysis) for breathable air and propellants. Conceivably, such propellants could be used for ascent stages and also in vehicles designed to 'ballistically hop' across the lunar surface to support exploration. Iron oxide can be found at all latitudes and may be reduced to iron, a component of many propellants and useful for construction. Indeed, the commercial exploitation of lunar resources is coming within reach, as we will discuss in Chapter 9.

In 2019, US President Trump committed NASA to return to the Moon by 2024. Given the timescale involved, it is clear that the next few years will be interesting and crucial for the human exploration of space, with the Moon being a key focus.

ENDNOTES

1. Its original name was the Lunar Excursion Module or LEM, later changed to LM (still pronounced LEM). Anecdotally, the change came about because an 'excursion module' sounded like something used on a family trip to the beach. Similarly, common astronaut usage truncated the command and service module combination (CSM) to just command module (CM).
2. A preliminary inspection was made during the translunar coast.
3. For later landings, tracking of the radio signals from the LM allowed trajectory updates to be relayed.
4. This incident is a testimony to the exceptionally robust nature of the code programmed into the NASA computer system that included a complete set of recovery programmes that in this case cleared the low priority jobs and re-established the important ones.
5. Various possibilities were considered, for example, the steady sinking of the landing legs into the surface.
6. One might have thought that the Moon was also the largest object in the night sky. However one of this book's reviewers pointed out that some faint (but visible to the naked eye) galaxies cover a larger area of the sky than the Moon does.
7. The orbit was set 69 miles above the surface. The joke among the astronauts involved in running simulations for training purposes was 'Wait until you see the 70-mile mountain on the far side!'
8. Public outcry at this treatment was apparently so great that Verne had to write a sequel titled *Around the Moon* in which the characters made a safe return journey.
9. The centre of gravity of an object is its point of balance. If you balance an object on your finger, the centre of gravity must lie on a vertical line through your finger. Imagine that I make a model of the Moon by taking two identically sized spheres and cutting them in half. One sphere is made from metal and the other from wood. If I join them to make a single sphere, the geometrical centre would be the centre of the sphere and the centre of gravity would be nearer the heavy half.
10. Organic compounds may not necessarily indicate life. Organic chemistry is a branch of science dealing with carbon compounds. Many carbon compounds are found in living systems, but they do not have to be. Rudimentary compounds from which the complex molecules needed for life can be constructed have been found in gas clouds in space, in the atmosphere of Jupiter and in meteorites.
11. Do you mean that you have not tried this?
12. Other factors also come into account such as the angular momentum of the two bodies. Angular momentum is associated with rotational motion.
13. Canup R. M. 23 November 2012. Forming a Moon with an Earth-Like Composition via a Giant Impact. *Science*. 338(6110): 1052–1055. DOI: 10.1126/science.1226073.
14. Pieters, C. M.; Goswami, J. N.; Clark, R. N. et al. 2009. Character and Spatial Distribution of OH/H2O on the Surface of the Moon Seen by M3 on Chandrayaan-1. *Science*. 326(5952): 568–572.
15. https://www.jpl.nasa.gov/news/news.php?feature=7218

2

The best driver in physics

2.1 THE FIRST VOYAGE TO THE MOON

The Apollo 8 mission was one of the most ambitious that NASA has ever undertaken. The crew originally trained for an Earth orbit test of the lunar module, but delays completing that complicated craft left room for a change in plans.

The CIA uncovered intelligence suggesting that the Russians intended to send one of their manned *Soyuz* spacecraft to the Moon. Keen and ambitious engineers at NASA wanted to ensure that the Russians did not reach another milestone before the American space programme. They lobbied to send the second manned Apollo to orbit the Moon – not just to loop around and return as the Russians planned. Consequently, on December 21, 1968, Frank Borman, William Anders and James Lovell lifted off for the Moon on the first manned launch of the Saturn V booster. The mission was a spectacular success, full of history-making events. The people who watched the live television transmission on Christmas day 1968 will never forget hearing the astronauts reading from the Book of Genesis while the pictures showed the passing lunar surface and the Earth rising over the Moon[1] (Figure 2.1).

The equipment performed faultlessly and the flight home became almost boring for the three-man crew. At one point, Mike Collins (later to be lunar module pilot for Apollo 11) was talking to the crew from the ground and mentioned that his son was wondering who was driving 'up there'. Anders dry reply was 'I think Isaac Newton is doing most of the driving right now.'[2]

2.2 ON SPACECRAFT AND SHOPPING TROLLEYS

One might think that the business of flying a spacecraft to the Moon was terribly difficult, requiring an astronaut to have his or her hands 'on the wheel' at all times. However, space flight is, in some ways, a good deal simpler than most people imagine. This is because the events that we see and experience around us govern our imaginations. In my experience, objects frequently do not travel in the directions that I intend: I struggle with shopping trolleys, for example.

Partly this is due to the abused lives that shopping trolleys lead. Most of them have four wheels that point in different directions. However, a lot of the problem is *friction*. The constant force of friction distorts our thinking about how objects move. I suspect that most people believe that you have to continually push an object to keep it going as you must with shopping trolleys. Similarly, I imagine that a lot of people think that Apollo's engines were continually burning, otherwise the spacecraft would grind to a halt (quite a lot of careless science fiction gets this wrong[3]). However, space contains no frictional forces and this makes a huge difference, so our intuition of how things ought to move can lead us astray. Newton imagined what it would be like without friction and from this picture deduced the laws of motion that dictate how objects really move. Newton's laws are clearly visible in the context of space.

2.3 THE POWER OF IMAGINATION

Newton would have understood all of the physics involved in navigating to the Moon and back. His laws of mechanics, first published in the 1600s, are perfectly adequate to calculate the orbits and trajectories required to navigate about the whole solar system.

In his book, *Principia Mathematica*, Newton described the principles that would guide all aspects of physics right up to the start of the 20th century. *Newtonian mechanics* was eventually replaced by the *theory*

FIGURE 2.1 Earthrise over the Moon. This is the first image of Earthrise taken by Bill Anders of the Apollo 8 crew. The original image was black and white, but modern processing technology has been applied to this image which can be viewed in colour at https://apod.nasa.gov/apod/ap181224.html. (Image courtesy of NASA, Apollo 8 Crew, Bill Anders. Processing and license courtesy of Jim Weigang.)

of relativity and by *quantum mechanics* between 1905 and 1930. However, relativity is only noticeably different from Newton's theory when an object is moving at a speed close to that at which light travels: 300 million metres per second.[4] No spacecraft ever moved faster than a small fraction of this speed. Similarly, quantum theory shows the limitations of Newton's mechanics when the objects involved are very small, typically the size of atoms, about 10^{-10} m across. Spacecraft do not fall into either of these categories.

The cornerstones of Newtonian mechanics are the three *laws of motion*, first set out in their completely correct form in the *Principia* (Galileo had been close to formulating the same ideas, but lacked Newton's mathematical insight). The first two laws form a pair that describe how objects move with or without forces acting on them. The third law tells us how forces and objects interact with each other.

Newton's laws of motion

- If an object does not have a net force applied to it, the object will remain stationary; if it was moving, it will continue to move in a straight line at a constant speed.
- If an object does have a net force applied to it, the object's motion will change in direction or speed or both.
- If object A applies a force to object B, then object B will also apply a force to object A and this force will be the same size as the first one, but in the opposite direction.

These laws seem so simple, it is a wonder that it took so long for someone to figure them out. The problem is that the first law in particular is contrary to the naïve observations of everyday life. A shopping

The best driver in physics

trolley pushed along and then released will roll across the floor, gradually slowing down until it comes to rest. Observations like this (but not with shopping trolleys in Newton's time) encouraged people to think, mistakenly, that force was required to *keep* an object moving. It took Newton's powerful imagination to extract the truth from the observations of daily life. If an object moved through empty space with no forces acting on it, there would be nothing to stop the object; it would coast along at a constant speed. Furthermore, if an object was subject to a combination of forces that cancelled each other out, then it would act as if no force was present. For example, if two equal forces pulled on a stationary object from opposite directions, the object would stay exactly where it was.

The phrasing used in the laws of motion caters for this circumstance with the term '*net force*'. The net force on an object results from adding all the forces acting on it, taking both their sizes and directions into account.

If an object simply floated without moving in any direction, then there would be no reason for it to start moving. Something would have to push it to get it going (a net force), but once it was moving the force would not be needed anymore. A shopping trolley in space could be pushed once, and then it would serenely coast along at a constant speed in the direction it was pushed. We will see in Chapter 4 that Newton was also able to visualise how an object could stay in space and permanently orbit the Earth.

The second law is often written in the form of a mathematical equation:

$$\text{Force} = \text{Mass} \times \text{Acceleration}$$

where acceleration is the rate at which the velocity[5] is changing:

$$\text{Acceleration} = \frac{\text{Change in velocity}}{\text{Time elapsed}}$$

For example, my favourite car, the Aston Martin DB11,[6] can accelerate from 0 to 62 mph in about 3.8 s or:

$$\text{Acceleration} = \frac{62 - 0 \, \text{mph}}{3.8 \, \text{s}} = \sim 16 \, \text{mph per second}$$

In other words, the speed of the car increases by 16 mph every second while it is accelerating. However, physicists prefer to work in metres so that speed is measured in metres per second (m/s). Using these units, the calculation is:

$$62 \, \text{mph} = 99.8 \, \text{km per hour} = 99{,}800 \, \text{m per hour} = 99{,}800 \, \text{m per 3600 seconds}$$

$$\text{so} \qquad 62 \, \text{mph} = \frac{99{,}800 \, \text{m}}{3{,}600 \, \text{s}} = 27.7 \, \text{m/s}$$

$$\therefore \text{Acceleration} = \frac{27.7 - 0 \, \text{m/s}}{3.8 \, \text{s}} = 7.3 \, \text{m/s}^2$$

Hence, the speed of the car increases by 7.3 metres per second every second. Physicists like to write this as 7.3 m/s². Note that this does not mean that we are dividing by a 'square second' (whatever that is!), although the result is pronounced 'metres per second squared' – which adds to the confusion.

In contrast to the Aston Martin, the Saturn V accelerated off its launch pad at 1.92 m/s² which seems leisurely in comparison. All the images we see of rockets lifting off suggest that vast forces are at work and surely these forces cannot result in something crawling off the launch pad. Two issues are involved. First, the rocket consumes fuel as it accelerates and gets progressively lighter. As a result, the acceleration increases. By the time the first stage completed its burn, the Saturn V was accelerating at something like 25 km/s^2. Second, given the vast mass of the rocket at launch, it was an achievement to produce even this initially sluggish acceleration.

Newton's second law allows us to calculate the force required to achieve a given acceleration:

$$\text{Force required} = \text{Mass of object} \times \text{Acceleration required}$$

In this formula, if we use mass in kilograms (kg) and acceleration in metres per second squared (m/s^2), the force is expressed in Newtons (N) – units of force named after Sir Isaac Newton. To get a feel for the size of a 1 N force, remember that a 1 kg mass has a weight[7] of 10 N.

We can now calculate the force needed to lift a Saturn V booster off the ground. As it takes off, the rocket is not moving fast enough for air resistance to be a problem (although resistance has a big effect later in the flight), but the engines have to overcome another force, which is considerable. Saturn Vs lifted off vertically, so the full weight of the rocket was acting downward, opposite to the direction in which it was trying to accelerate. In order to move even a centimetre off the ground, the engines had to develop a force at least as big as the weight of the rocket,[8] that is, 28,400,000 N.

When fully laden with fuel, the Saturn V had a mass of 2,900,000 kg (2,900 tonnes[9]). To accelerate off the launch pad at 1.92 m/s^2 the net force needed to be:

$$(\text{Force of engines} - \text{Weight of rocket}) = (2{,}900{,}000 \text{ kg}) \times (1.92 \text{ m/s}^2) = 5{,}570{,}000 \text{ N}$$

making the force of the engines:

$$28{,}400{,}000 \text{ N} + 5{,}570{,}000 \text{ N} = 34{,}000{,}000 \text{ N} \text{ (3 significant figures)}$$

In contrast, the force of the very nice V12 petrol engine installed in the Aston Martin DB11 is:

$$\text{Force of DB11 engine} = \text{Mass of car} \times 7.3 \text{ m/s}^2 = (1{,}875 \text{ kg}) \times (7.3 \text{ m/s}^2) = 13{,}700 \text{ N}$$

Alternatively, if the DB11 could muster the same force as the Saturn V, its acceleration off the line would be:

$$\text{Acceleration} = \frac{\text{Force}}{\text{Mass}} = \frac{34{,}000{,}000 \text{ N}}{1{,}875 \text{ kg}} = 18{,}000 \text{ m/s}^2$$

which is very brisk ...

2.4 FALLING

At the end of the last Apollo 15 moonwalk, astronaut David R Scott carried out a short demonstration for the TV audience. The experiment is familiar to physics teachers, but he had the advantage of being on the Moon, where there is a near-perfect vacuum. Scott dropped a falcon feather (the lunar module for that mission was named *Falcon*) and a geological hammer from the same height (Figure 2.2). Despite their obvious differences in weight, they struck the lunar surface at the same time.

The pull of gravity on the Moon is less than that on Earth, but the physics is the same: all objects accelerate under the action of gravity at the same rate. On Earth, air resistance masks this effect. Try dropping a pen and a piece of paper from the same height. The pen will thud onto the floor, while the paper flutters down slowly. Now take the paper and squeeze it into a ball. Dropping the paper ball and the pen gives a different result: they should hit the ground at pretty much the same time. However, the mass of the paper has not changed; the air resistance acting on the paper has been greatly reduced.

FIGURE 2.2 Dave Scott on the surface of the Moon with a hammer in his right hand and a falcon feather in his left. On live television, he demonstrated one of Galileo's important discoveries: all objects fall at the same rate under the action of gravity. On Earth, air resistance in the atmosphere makes this concept more difficult to show. (Image courtesy of NASA.)

When an object is released, it accelerates towards the ground as the planet pulls on it with the force of gravity, the force we call the *weight* of the object. The size of this force is determined by the mass of the object: the greater the mass, the heavier the weight:

$$\text{Weight} = \text{Mass} \times \text{Gravitational field strength} = \text{Mass} \times g$$

where g represents the gravitational field strength acting on the surface of the planet in Newtons per kilogram (N/kg). On Earth, this happens to be $\sim$9.8 N/kg. The strength of gravity is down to the mass (and radius) of the planet. On the Moon, the strength of gravity is $\sim$1.63 N/kg, that is, $1.63/9.81 = 0.166$ or $\sim$1/6 the strength on Earth; hence we often read that the gravity on the Moon is 1/6 g.

Newton's second law of motion tells us that the *larger* the mass of an object, the *smaller* the acceleration produced by a given force.

If we combine these two pieces of physics, something extraordinary results. When an object falls and we can ignore air resistance, the only force acting is its own weight (technically this is called *free fall*). To determine the resulting acceleration, we calculate as follows:

$$\text{Acceleration while falling} = \frac{\text{Force}}{\text{Mass}} = \frac{\text{Weight}}{\text{Mass}} = \frac{mg}{m} = g$$

The resulting acceleration does not depend on the mass at all and has the same numerical value as the strength of gravity.

To 'see' this physically, compare two falling objects. The one with a small mass will have a small force acting (its weight), but a small force can still produce quite an acceleration on an object with little mass. The heavier object has a greater force acting on it, but as it is a larger mass, this greater force will not produce an acceleration exceeding that experienced by the small mass falling next to it.

The familiar acceleration due to gravity on Earth (abbreviated as g) gives us a convenient way of scaling large accelerations (such as those experienced by astronauts during liftoff). For example, an acceleration of $10g$ would be 98.1 m/s^2.

The force required to produce an upward acceleration of $10g$ is 11 times the weight of an object. To achieve that acceleration, the force must overcome the weight of the object and then produce an upward acceleration of $10g$, so:

$$(\text{Force} - \text{Weight of object}) = \text{Mass} \times \text{Acceleration} = m \times 10g$$

$$\therefore \text{Force} = m \times 10g + \text{Weight of object} = 10mg + mg = 11mg = 11 \times \text{Weight of object}$$

Astronauts experiencing $4g$ during launch feel five times heavier due to the acceleration.[10] Fighter pilots can 'pull' up to $8g$ when they bank their aircraft into a tight turn or roll. Formula 1 cars often generate accelerations of $6g$ when they turn corners.

An important practical upshot of this property of gravity, that all objects fall at the same rate, is that astronauts in a spacecraft orbiting the Earth feel 'weightless'[11] as they fall towards the Earth with the same rate of acceleration as the spacecraft they occupy. How orbits allow objects to fall without hitting the ground is covered in Chapter 4.

2.5 FORCES DURING LIFTOFF: THE ASTRONAUTS

Anyone who has seen a film of astronauts during liftoff will recall that they were 'pressed down' into their acceleration couches and perhaps the skin on their faces was distorted and 'pulled down' over their cheekbones. Sometimes this is used as artistic licence in science fiction films to make a scene more dramatic, but films of real events show that the basic effects are present when a large rocket booster takes off.

Even though astronauts describe the forces they experience at launch as being 'pressed into couches by giant hands', careful analysis shows that the effect is actually due to an *upward force* pushing them off the ground.

Figure 2.3 shows a stylised astronaut lying on an acceleration couch. The forces are indicated as arrows. Note that only two forces act directly on the astronaut: his weight acting downwards and the upward force applied by his couch. The rocket engines do not exert a force directly to his body because he is not directly connected to the engines.

The astronaut is strapped into an acceleration couch, which is in turn (presumably) bolted to the capsule, so the couch experiences a force from the engines transmitted via the body of the spacecraft. Due to the action of this force, the couch will start to accelerate upward. The astronaut, however, does not

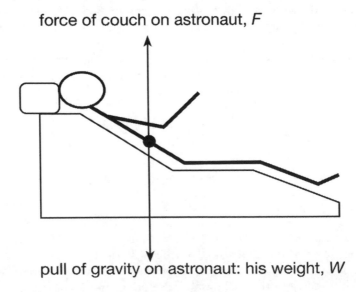

FIGURE 2.3 The forces acting on an astronaut during liftoff.

immediately accelerate with it. At the moment that the rocket starts to move, the size of the force that the couch exerts on the astronaut is just equal to his weight, otherwise he would be falling through the couch. As the rocket lifts off, *the couch accelerates past him*. The padding in the couch is distorted and he appears to sink into it, not because he is being pushed down, but rather because the couch is trying to overtake him! As the material distorts further, the force that it applies to the astronaut will increase *until it is sufficient to make him accelerate upward at the same rate as the couch*. At that point, he will stop sinking further into the couch while the compression in the padding remains constant.

The same effect causes human faces to be distorted. The skin of the face is not rigidly attached to the bone beneath, so the head tries to accelerate through the skin causing a 'drawn over' effect.

If this still does not seem plausible, imagine an astronaut standing on a weighing machine (Figure 2.4) rather than lying on a couch. Weighing machines work by measuring the contraction of a spring when someone stands on it. The heavier the person, the greater the compression of the spring. As the spring contracts, the force that it applies upward to support the person increases. The compression stops when the upward force is equal to the person's weight.

As a rocket starts to accelerate, the force acting upward on the spring increases but the force acting downward remains the same as it is due to the weight of the astronaut. In fact, this force never changes, so the upward force exceeds the downward force, which is why the spring starts to accelerate upward.[12] The astronaut at first does not accelerate as the two forces exerted on him are still equal. This means that the bottom of the spring moves upward while the top is stationary, which causes the spring to shorten. As the spring compresses more, the upward force that it exerts on the astronaut increases. Now the forces on him are no longer equal and he also starts to accelerate upward. The spring continues to compress until the net force acting on the astronaut is enough to make him accelerate upward at the same rate as the entire rocket structure. From that point onward, the two ends of the spring move in tandem.

The end result is that the weighing machine records a greater value (as the spring is now compressed more), yet the force of gravity has not changed. We will return to this when we discuss weightlessness in Chapter 4.

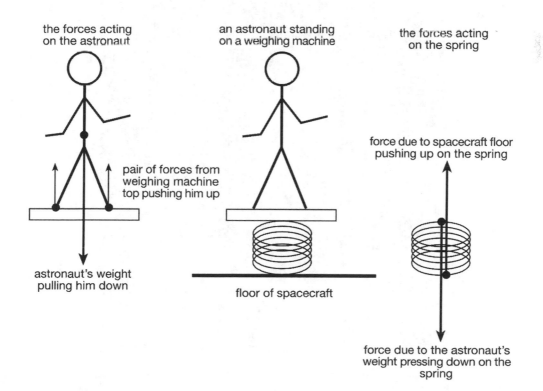

the forces acting
on the astronaut

an astronaut standing
on a weighing machine

the forces acting
on the spring

pair of forces from
weighing machine
top pushing him up

astronaut's weight
pulling him down

floor of spacecraft

force due to spacecraft floor
pushing up on the spring

force due to the astronaut's
weight pressing down on the
spring

FIGURE 2.4 The forces acting on a weighing machine.

The astronaut's couch behaves just like the spring. In fact, all objects behave a little like springs. When you walk on a floor, it distorts a tiny amount beneath you to resist your weight. A plank resting across two oil drums will bend in the middle to support someone standing on it. The same idea applies to an astronaut's couch.

During liftoff, astronauts report feeling heavy and say that it is much harder to lift their arms to operate the switches on the control panel above them. During liftoff they feel heavier; in orbit, they feel weightless.[13]

Consider raising an arm to press a switch on a control panel (Figure 2.5).

To lift an arm off his couch, an astronaut must make the arm accelerate slightly faster than the rocket. If he fails to achieve that, the couch will catch up with the arm and from the astronaut's view, his arm will fall back. To reach the necessary level of acceleration, an astronaut must apply an internal muscular force much greater than the weight of his arm as shown in Figure 2.6.

If a spacecraft accelerates upward at $4g$, we can calculate the upward force on an arm held in the air and not resting on the couch:

$$\text{Mass of arm} \times 4g = (\text{Upward force from muscles} - \text{Weight of arm})$$

$$\therefore 4\,mg = F - mg$$

so:

$$F = 5mg$$

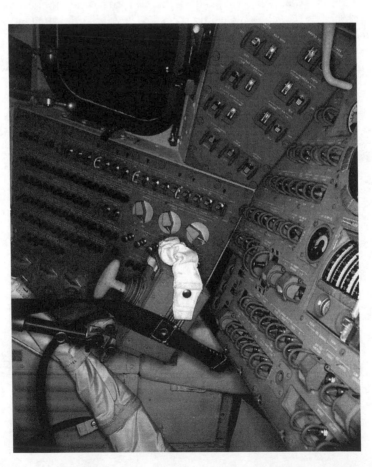

FIGURE 2.5 An interior view of the Apollo 16 command module during post-flight inspection. The control panel spanning the top of the interior can be seen. During launch, the crew members lay on their backs below the control panel. (Image courtesy of NASA; scan courtesy of P. L. Pickering.)

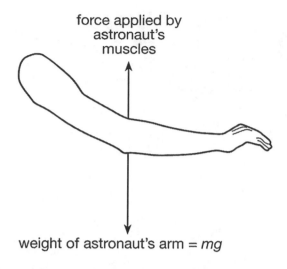

force applied by
astronaut's
muscles

weight of astronaut's arm = mg

FIGURE 2.6 The forces acting on an astronaut's arm.

The astronaut has to apply a force five times greater than the weight of their arm, just to hold that arm in the air. The force would have to be a little bigger to lift the arm towards the control panel.

2.6 FORCES DURING LIFTOFF: THE SPACECRAFT

Figure 2.7 shows the launch of Apollo 15, with the first stage of the Saturn V powering the spacecraft off the pad. The bright column of burning exhaust from the engines strikes the ground and seems to spread in all directions. Actually, a significant portion of the billowing cloud surrounding the launch tower is steam

FIGURE 2.7 The launch of Apollo 15. (Image courtesy of NASA.)

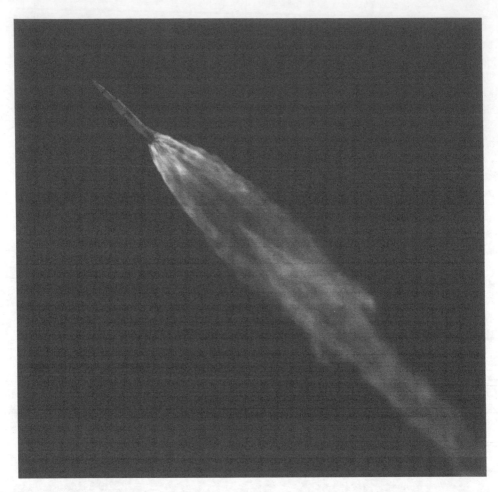

FIGURE 2.8 Apollo 10 in flight. The mission took astronauts Eugene Cernan, John Young and Thomas Stafford to the Moon on a test flight of the CSM and the lunar module. (Image courtesy of NASA.)

produced by 106,000 litres of water per minute pumped into flame buckets designed to absorb the rocket exhaust under the launch tower.

Images like this conjure the impression that the rocket is lifted by the action of the exhaust gases pushing against the ground, rather like a person getting out of a chair by pushing down onto it. However, if you consider Figure 2.8, showing a Saturn V in flight, the rocket is clearly accelerating through the upper atmosphere and will eventually reach space without pushing against anything.

Interestingly, the 'pushing against' thinking is so natural, even Robert Goddard, the pioneer of practical rocketry, had to contend with it. In a series of experiments during the 1920s, he established the principles on which boosters such as the Saturn V were developed. However, much of his work was not appreciated at the time. In a famous editorial, the *New York Times*[14] pronounced:

> That Professor Goddard, with his 'chair' in Clark College and the countenancing of the Smithsonian Institution, does not know the relation of action to reaction, and of the need to have something better than a vacuum against which to react – to say that would be absurd. Of course, he only seems to lack the knowledge ladled out daily in high schools.

The editor's comment about action and reaction refers to Newton's third law of motion. As a simple example of this law at work, consider what happens when you rise from a chair. Your first move when standing up is to apply a force to the chair (*action force*): you put your hands on the arms and push *down* but your body moves *up*. It is *not* the force that you apply to the chair that moves you. Newton's third law of motion tells us that the chair, in response to the force that you apply, pushes *back* on you (*reaction force*). It is this force, the force of the chair on your body, that gets you moving.

Similarly, if you set out to walk across a room, what gets you moving is not a force you apply to your body. As you start walking, your leading foot lifts, moves forward and makes contact with the floor. At this point, you push *backward* with that foot. The *floor* then pushes *forward* on your foot and that force propels you.

Imagine hanging from the ceiling by a string. All the pushing and waving of feet and arms that you can muster will not move you backward or forward to any extent because you are not in contact with anything. Your limbs cannot apply a force to an object (like a floor or a chair) to trigger a reaction force (Newton's third law) to your body. Astronauts carrying out space walks face the same problem. When they float freely in space, they have no object to push in order to trigger a reaction force on them.

Pictures of liftoff seem to show the rocket pushing on the ground, and so we imagine that the rocket is set in motion when the ground pushes back. This is why people such as the editor of the *New York Times* thought that rockets could not work in space because they would have nothing to push against.

Newton's third law applied correctly to a rocket deals with the exhaust and the rocket *body* not the exhaust and the ground. The physics is not straightforward and is more easily understood by applying the concept of *momentum*.

2.7 MOMENTUM

Nowadays, we are accustomed to seeing pictures of astronauts working in space. We have watched the construction of the International Space Station and the repair of the Hubble Space Telescope, both of which required practical engineering in space. Successes like this tend to make the process of working with tools and objects in free fall situations seem very easy. In reality, this work is far from easy and many of the tools must be specially designed.

On Earth, the procedure for undoing a screw is straightforward.[15] You apply force to a screwdriver, which in turn applies a force to the screw which loosens it. The screwdriver works like a lever so that the force it applies to the screw is greater than the force you apply to the screwdriver. However, this apparent simplicity hides the vital role that friction plays in the process. In space, where there is no air resistance and an astronaut may not have a convenient surface to press against, Newton's third law has some awkward consequences. If an astronaut applies a force to a screwdriver and hence to a screw, a force will be applied back on them. The force that turns the screw one way triggers a reaction force that turns the astronaut in the opposite direction. On Earth, a person using a screwdriver is in contact with the ground and the force of friction transfers the tendency to rotate through to the Earth itself, where it has a negligible impact.

Tools for use in space have to be specially designed to absorb the reaction effect. However, this is not the only hazard that has to be overcome. The apparent weightlessness of objects could lead to mistakes. On Earth, a large box of electronics is both heavy (has a large *mass* so that the Earth pulls on it with a large gravity force) and awkward to move (its large mass requires a large force to achieve acceleration). The same box in space appears to have no weight; it floats. Anyone trying to push it along in space gets quite a surprise. The box still has a large mass and requires a large force to get it moving. Unfortunately, stopping it also requires some force!

In both cases, applying the necessary force to the box will cause the box to exert an equal force back (Newton's third law again). If the astronaut has less mass than the box, this force will have a greater effect on the astronaut than the astronaut had on the box.

Imagine that astronaut A pushes a massive box of electronics towards astronaut B. As a result of A's application of force to the box, the box pushes A back and A drifts away from the box. As the box arrives at B, she extends her hands to slow the movement of the box. The force she applies slows the box down, but unfortunately the force the box applies to her speeds her up. The result is that both the astronaut and the box start moving in the same direction. Sometimes the laws of physics hurt.

The situation of the two astronauts and the box can be fully analysed by considering the *momentum* of each object. Momentum is a useful quantity that often allows physicists to analyse a situation that involves very complex details.

The momentum of an object is defined as its mass multiplied by its velocity:

$$\text{Momentum} = \text{Mass} \times \text{Velocity}$$

$$P = mv$$

so any object that is not moving has zero momentum. Newton's laws can be used to prove that the total amount of momentum in the whole universe never changes. If we were able to add the momenta of every single object in the whole universe now and be able to calculate the same impossible sum a million years from now (or a million years ago for that matter), you would get exactly the same answer.

One of the remarkable things about Newton's laws is that they allow us to make statements like this without even being able to add up all the momenta in the universe. What we can do is investigate what happens whenever objects collide. Our example of the two astronauts and the box is exactly the right sort of situation.

At the start of the example, the astronauts and box are not moving so their total momentum is zero.

Assume that astronaut A pushes on the box with a force of 50 N for 0.1 s. The box has a mass of 200 kg, so it will accelerate at the rate of:

$$\text{Acceleration} = \frac{\text{Force (N)}}{\text{Mass (kg)}} = \frac{50\,\text{N}}{200\,\text{kg}} = 0.25\,\text{m/s}^2$$

An acceleration of 0.25 m/s² for 0.1 s results in an increase in speed:

$$\text{Acceleration} = \frac{\text{Change in speed}}{\text{Time}}$$

$$\therefore \text{Change in speed} = \text{Acceleration} \times \text{Time} = (0.25\ \text{m/s}^2) \times (0.1\ \text{s}) = 0.025\ \text{m/s}$$

As this is a change in speed and the box originally was not moving, we conclude that the box will coast away from astronaut A at a gentle speed of 0.025 m/s.

Now consider what happens to astronaut A. He is also acted upon by a force of 50 N, but the difference is that he and his spacesuit have a total mass of 100 kg. As a result, he accelerates at a rate of 0.5 m/s² for 0.1 s and moves away from the box at a speed of 0.05 m/s.

Now that we know the velocity of the box and the velocity of astronaut A, we can look at the total momentum:

$$\text{Momentum of box} = \text{Mass} \times \text{Speed} = (200\ \text{kg}) \times (0.025\ \text{m/s}) = 5\ \text{kg m/s}$$

$$\text{Momentum of astronaut A} = \text{Mass} \times \text{Speed} = (100\ \text{kg}) \times (0.05\ \text{m/s}) = 5\ \text{kg m/s}$$

At first, this does not look like it will add up to zero – the original total momentum. Then we realise that the box and astronaut A *move in opposite directions; perhaps we should treat one of them as a negative.* The choice of box or astronaut does not matter so we will take the box as positive:

$$\text{Total momentum} = \text{Momentum of box} + \text{Momentum of astronaut A} = (5\ \text{kg m/s}) + (-5\ \text{kg m/s}) = 0$$

The collision between the box and astronaut B is not so easy to analyse because we cannot be sure about the sizes of the forces. The result will depend on many details such as how astronaut B moves her arms as the box hits her (people often move their hands back as they catch a ball; this minimises the forces: the same principle applies when car designers make the fronts of cars crumple on impact). However, we can use momentum principles to find the result.

From our previous calculation, the box coasts along with a momentum of 5 kg m/s before it reaches astronaut B. After the collision between the box and the astronaut, they drift along together, almost certainly at different speeds. It would be hard to calculate the two speeds accurately without knowing the details of how the box and astronaut came together and then moved apart. However, we can say that the total momenta of astronaut B and the box after their collision must add to 5 kg m/s. If that is not true, the total momentum in the universe would have changed when the box and astronaut collided. This is the power of momentum analysis. It enables us to study the end result of a collision or interaction between objects without necessarily understanding the details. In this case, once we are confident that the total momentum does not change, then we can certainly calculate the final momenta of the box and astronaut.

There are many situations in which momentum does not seem to be a conserved quantity. For example, consider a person who suddenly decides to start walking. The person's momentum increases without any obvious decrease of momentum for another object. In fact, there is a corresponding decrease. In order to start moving, the person applies a force to the ground. This force alters the momentum of the Earth. In response, the Earth pushes back, altering the person's momentum. There is a transfer of momentum from one to the other. Of course, the speed change of the Earth is too tiny to measure because the Earth is so much more massive than a person.

2.8 THE PHYSICS OF ROCKET MOTORS

We can gain some insight into how rocket motors provide propulsion by considering the following situation. An astronaut finds herself adrift, having accidentally let go of her spacecraft during a spacewalk. In reality, this is not a realistic situation because safety precautions require tethering of astronauts to the spacecraft hull.[16] After some experimentation, she finds that she is unable to make any progress towards the spacecraft by waving her arms, kicking her legs or making swimming motions. She cannot apply a force to any object and so is unable to trigger a reaction force that would get her moving. Fortunately, she is well trained in the laws of physics. She removes a large spanner from her tool belt and throws it away as hard as she can (but not at the spacecraft!). As a result, she starts to drift through space in the opposite direction.

We can see that the initial momentum of the spanner plus astronaut was zero (neither was moving). As a result of the astronaut's throw, the spanner gained momentum. To conserve the initial zero momentum, the astronaut also gains momentum, but in the opposite direction. The forces involved in this case are quite simple. The astronaut applies a force to the spanner; by Newton's third law the spanner applies a force back and that force sets the astronaut moving.

An astronaut stranded in space probably has only one spanner. Although she could remove other items and throw them away (up to a point), it would be far better to provide her with a continual source of mass that she can use to produce propulsion. Let's give her a large balloon inflated to a high pressure.[17] (If this was a science fiction film, we could heighten the tension by having her use the hose that supplied oxygen to her spacesuit.) By carefully opening the end of the balloon, she can release some of the gas inside and so propel herself across space. With a certain amount of ingenuity, she can even use the balloon to steer by pointing it in various directions.[18] Although intuition tells us that this idea will work, we need to consider the physics involved.

From a momentum view, this scenario seems quite simple. The initial momentum was zero. The escaping gas gains momentum in one direction and the astronaut gains equal momentum in the opposite direction. But this is not quite right. How did the escaping gas gain momentum? Opening a hole in the side of the balloon does not apply a force to the gas to give it momentum. It makes more sense to think that the escaping gas already had momentum inside the balloon.

The balloon initially has zero momentum, even though all the gas molecules inside move about quickly because there is as much momentum in one direction as there is in the opposite direction at any moment. The gas molecules move randomly inside the balloon, which ensures that any unevenness in their motion cancels out very quickly. The internal momenta always add up to zero. Consider a simple case with only

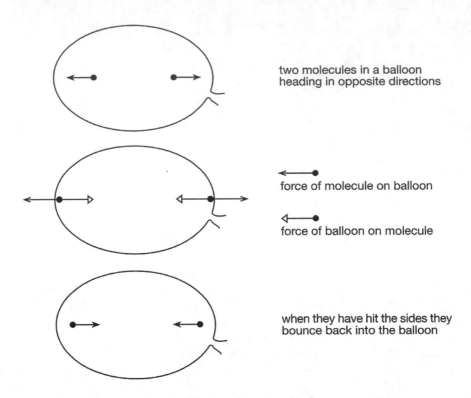

two molecules in a balloon
heading in opposite directions

force of molecule on balloon

force of balloon on molecule

when they have hit the sides they
bounce back into the balloon

FIGURE 2.9 Two molecules bouncing back and forth inside a closed balloon.

two molecules (Figure 2.9). One is moving towards the left, the other towards the right. When the molecules encounter the far ends of the balloon, they bounce off the skin and head back in the opposite direction. This back-and-forth movement continues and always maintains zero momentum.

A molecule applies a force to the skin of a balloon when it strikes the skin. By Newton's third law, the skin applies an equal force back and the molecule reverses direction. The skin is pushed out in the direction in which the molecule first moved. The same effect occurs at the other end of the balloon. The skin at that end is also pushed. This is why the balloon does not move: it is pushed equally in both directions (on average) by the molecules striking its skin.

However, if we remove one side of the balloon (Figure 2.10), or more realistically untie the knot keeping the balloon sealed, the result is somewhat different. The molecule heading to the right (in Figure 2.10) does not bounce off skin; it flies out of the balloon. As a result, the molecule does not apply a force at that end. Now there is nothing to balance the force produced by the molecule hitting the other end of the balloon. This unbalanced force causes the balloon to move forward (left) as the gas escapes from the other end. This is the secret of rocket propulsion.

We must add one more component to turn our balloon example into a device capable of sustaining thrust for an extended period: the generation of pressure without the need to inflate a balloon. We can imagine a cartoon rocket powered by a large balloon pumped up on the launch pad, but such engineering would be problematic at best.

In a practical rocket, two chemicals react to provide a great deal of heat.[19] One of the chemicals acts as a *fuel* and the other as an *oxidiser*. When wood burns in air, the heat triggers a reaction between chemicals in the wood (fuel) and oxygen in the air (oxidiser). If a rocket is to work high in the atmosphere and in space, it must carry the required oxidiser with it. The Saturn V first stage motors used liquid oxygen with kerosene as the fuel. A secondary chemical ignited spontaneously on contact with the air to trigger combustion of the main reactants (Figure 2.11).

The fuel and the oxidiser are mixed in a *reaction chamber* that plays the role of the balloon. Combustion in the chamber produces exhaust gases with very high energy: in other words, it yields molecules that move at very high speeds. In essence, combustion produces a high pressure similar to the pressure within the

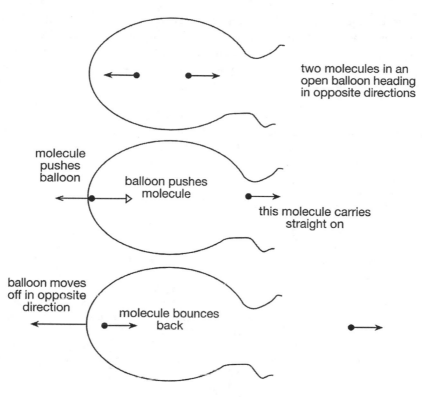

two molecules in an
open balloon heading
in opposite directions

molecule
pushes
balloon

balloon pushes
molecule

this molecule carries
straight on

balloon moves
off in opposite
direction

molecule bounces
back

FIGURE 2.10 Molecules in an open balloon.

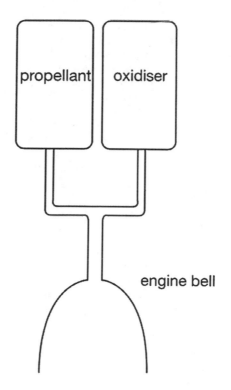

propellant   oxidiser

engine bell

FIGURE 2.11 The basic design of a chemical rocket involves the reaction of a fuel with an oxidiser, both of which are stored in tanks inside the vehicle.

balloon, but this process is sustainable, at least while chemicals are still present. At one end of the reaction chamber, a hole allows exhaust materials to escape. The pressure inside the reaction chamber becomes unbalanced, as occurs when propulsion results from the open balloon. We see the exhaust gases flooding spectacularly from the engine, which misleads people into thinking that the acceleration results from gases pushing against the ground. In reality, unseen gases pushing upward inside the reaction chamber produce the thrust.

The figures for the Saturn V's first stage are staggering. In flight, its five engines consumed more than 4,900 litres of kerosene and over 7,500 litres of liquid oxygen per second. The rockets generated temperatures in excess of 3,200°C and produced a total force exceeding 34,000,000 N.

ENDNOTES

1. I had the privilege of meeting Reginald Turnill (1915–2013), the iconic BBC commentator, whose voice was still very evocative of the Apollo era. When he came to talk to the school where I taught, I played a DVD of the television transmission from Apollo 8 and his emotion was still apparent after many decades.
2. In the *Apollo 13* movie, Tom Hanks (who played Jim Lovell) said 'we just put Sir Isaac Newton in the driver's seat' as the crew shut down the computers onboard. Lovell, who flew on Apollo 8, served as a consultant for the film and may have suggested the line.
3. There is no excuse for breaking the laws of physics just to create drama!
4. This comment specifically refers to Einstein's special theory of relativity. His general theory published 11 years later extended the special theory to include gravity. The theory further modified Newton's ideas, but becomes vital only when dealing with objects far more massive than the Sun.
5. Physicists like to distinguish speed from velocity. Velocity is the rate at which an object moves in a specific direction. An object moving in a circle could travel at a constant speed, but its velocity would change as its direction of motion alters while it moves around the circle.
6. In the first edition, I mentioned the Aston Martin DB7. Things move on …
7. Weight and mass are often confused. The mass of an object is difficult to define at an elementary level but is basically a measure of the amount of matter in an object. Conversely, weight is a force – the force with which Earth pulls on an object. Weight depends on the amount of matter an object contains and on the strength of the Earth's gravity. At the equator, the Earth pulls on every kilogram of mass with a force of 9.8 N.
8. There is no such thing as a typical Saturn V. The figure here was cited in the Apollo 11 press kit to indicate the mass of the rocket at first motion. This mass included the ice formed on the metal surface due to cold temperatures of the stored liquids. For this reason, the total is not the same as the sum of the various figures for stage masses and Apollo spacecraft masses quoted elsewhere in this book. It is very difficult to be consistent when discussing masses at various stages of countdown and launch.
9. The metric tonne (1,000 kg) corresponds to 1.1 US tons.
10. The peak acceleration experienced during a Saturn V launch was achieved just before the first stage was jettisoned and was about 4*g*.
11. Weightlessness can be a confusing term; it does not mean that an astronaut has no weight! Weight is the force of gravity exerted on an object. As long as an individual is near Earth (being in orbit is still near Earth), the force of gravity will act. See Chapter 4 for a further discussion of weightlessness in orbit.
12. In this account, we assume that the weight of the spring is negligible; otherwise that would be an additional force that would complicate but not invalidate the argument.
13. The force of gravity on an astronaut orbiting the Earth is slightly less than gravity on the ground. An astronaut in orbit is further from the centre of the Earth, but the gravity force in orbit is not zero. Weightlessness means the *apparent* absence of weight, not an absence of gravity.
14. The *New York Times*, January 13, 1920. The newspaper published a correction in 1969.
15. In principle the technique is straightforward. In practice, I have met many screws that seem to fight back.
16. On February 7, 1984, Bruce McCandless, one of the space shuttle astronauts, tested a rocket-powered backpack that allowed him to move about without being connected to the shuttle.
17. The balloon would have to be made of quite robust material to survive the pressure difference between the gas inside it and the vacuum of space outside it.
18. In *The Martian,* an excellent movie, a character cuts a hole in his spacesuit in order to use the escaping oxygen to thrust him towards rescue. In practice, it is likely that the pressure difference would force the palm of his hand against the hole, effectively sealing it.
19. Monopropellants are possible when a single chemical decomposes on contact with a catalyst.

3 Rocketry

3.1 FALTERING STARTS

Towards the close of World War II the German rocket scientist Wernher von Braun brought his team of engineers together in secret to discuss what they would do after the war. The Nazis had forced them to work on developing rockets (such as the V2) in the hope that they would be a last-minute weapon to defeat the allies. Von Braun dreamed of using their technology to explore space.[1] The group decided that their best chance of pursuing this goal was to surrender to the United States (they did not fancy going to Russia; they had seen enough of totalitarian regimes). Meanwhile, the Americans were combing occupied Germany for scientists who might be useful after the war ended. The Nazis instructed von Braun to destroy his papers (he hid them in an abandoned mine) and his team was shipped away for what they believed would be their execution. Fortunately, in the confusion during the last days of the war, they managed to surrender to the American forces.

However, when they arrived in America, they found a country that was not as keen to develop space rockets as they had hoped. The war ended, and Congress was too concerned with other matters to vote money for rocket and space research. Von Braun and his colleagues were posted to Fort Bliss in Texas to tinker with captured V2s and teach rocketry to interested members of the Army. This unfortunate situation continued for 6 years, during which the Soviets had their German scientists develop ballistic missiles.

By 1950, the US Army became convinced that Soviet rocket research was advancing, and so it mobilised von Braun. His team was moved to Huntsville, Alabama where they worked at Redstone Arsenal (a site the Army previously used to load explosives onto shells and bombs) on the development of missiles. Once again, they were obliged to construct weapons rather than peaceful rockets. Just weeks after they arrived, North Korea invaded the South and the word came down from President Truman: develop a ballistic missile capable of carrying a nuclear warhead 200 miles. Three years later, the first Redstone missile was launched from the military test range at Cape Canaveral in Florida.

By 1956, the US Air Force was developing the Thor, Atlas and Titan missiles (the latter pair were capable of firing warheads 5,000 miles) and the Army was working on its missile Jupiter. However, none were flight ready and the scientists needed to test the ability of the warheads to re-enter the Earth's atmosphere. Von Braun was asked to further develop the Redstone to carry out these tests. The team achieved this by lengthening the rocket, modifying the engines and adding two upper stages. The upgraded rocket was called Juno-C and it worked perfectly from the start.

Von Braun knew that adding a small further stage to the top of the stack would enable them to launch a satellite into orbit. They constructed the necessary equipment and asked for permission to proceed. The response from Washington was rather negative. The rumour was that the president was not about to allow the first US satellite to be launched by German scientists. Two years earlier he sanctioned the development of a non-military booster called Vanguard for precisely that purpose. Von Braun however, was worried. He realised that the Soviets were just as capable of adapting a ballistic missile to carry satellites as he was. He was convinced that Vanguard would not work in time. He was proven correct. On October 4, 1957, the Soviets put *Sputnik* into orbit while the modified Juno-C sat idly in a shed. Thirty days later, the Soviets launched a much heavier (500 kg) satellite with a live dog onboard.

Vanguard's first try, under enormous pressure from Washington, took place on December 6, 1957. The rocket was shipped to Cape Canaveral where, with extensive press coverage, it attempted to launch a 14 kg satellite. The launch was certainly spectacular. One of the commentators watching exclaimed, 'it was so quick, I really did not see it.' In fact, Vanguard exploded on the pad, blasting the satellite into the air to land in the surrounding scrub. Once there, it dutifully and plaintively started transmitting radio signals. Another member of the assembled press corps pleaded: 'why doesn't someone go out there, find it, and kill it?'

Finally, President Eisenhower relented and gave von Braun's team a chance. The modified Juno-C was taken out of storage, further changes were made and on January 31, 1958, it successfully placed the *Explorer 1* satellite into orbit. Onboard was a Geiger counter, developed by Dr James Van Allen, who consequently made the first scientific discovery of the new space age: Earth is surrounded by regions of radiation now known as the *Van Allen belts*.

Von Braun's Redstone would soon be used to launch the first American into space, but not before more political indecision and technical setbacks allowed the Soviets to get there first. Yuri Gagarin became the first human to orbit the Earth on April 12, 1961. Alan Shepard followed on May 5, 1961, although his much shorter flight was more of a sub-orbital ballistic hop. The space race was underway.

3.2 THRUST

Thrust is the term used for the force produced by a rocket motor. Applying Newton's second and third laws produces the following expression:

$$\text{Thrust (N)} = \text{Exhaust velocity (m/s)} \times \text{Fuel consumption rate (kg/s)}$$

$$T = u \times \frac{\Delta m}{\Delta t} \quad \text{where } u \text{ is the exhaust velocity}$$

Mathematicians use the Δ symbol to mean 'change in', so that Δm denotes a change in mass (not some quantity $\Delta \times m$) and Δt indicates a change in time or time elapsed. Hence, $\Delta m / \Delta t$ is the rate at which a mass changes. From this expression, we can see that maximising the thrust requires the exhaust gases to move at the highest possible speed and fuel consumption at a rapid rate. Calculating thrust starts by first noticing that the exhaust gas speed multiplied by the change in the propellant mass (i.e. mass ejected as exhaust) is the momentum of the exhaust:

$$\text{Mass of exhaust emitted in 1 s } (\Delta t = 1) = \Delta m$$

$$\text{Momentum of exhaust} = \text{Mass of exhaust} \times \text{Exhaust velocity} = \Delta m \times u$$

So, the expression for the thrust is related to the momentum of the fuel ejected as exhaust per second:

$$T = u \times \frac{\Delta m}{\Delta t} = \frac{u \times \Delta m}{\Delta t} = \frac{\text{Momentum of exhaust}}{\Delta t} = \frac{\Delta (mu)}{\Delta t}$$

The last part of this expression indicates that the thrust (which is the same as the force) is the change in momentum per second. This is another way of writing Newton's second law of motion and corresponds to the more general manner in which he devised it. In Newton's full form, the second law of motion reads:

$$\text{Force applied} = \frac{\text{Change in momentum}}{\text{Time}}$$

Later in this chapter, we will find that this is a more useful form of the second law when dealing with rockets.

3.3 PROPELLANT

A key factor in producing the highest possible thrust is the choice of propellant. Various chemical combinations have been developed for use in rocket motors, all of which have specific advantages and disadvantages.

Propellants are rated according to their *specific impulse*, I_{sp}, which is the thrust produced divided by the weight of propellant used per second:

$$I_{sp} = \frac{\text{Thrust}}{\text{Weight of propellant used per second}} = \frac{u \times \Delta m / \Delta t}{\Delta W / \Delta t} = \frac{u \times \Delta m / \Delta t}{g \times \Delta m / \Delta t} = \frac{u}{g}$$

showing that the specific impulse is the ratio of the exhaust speed to the strength of gravity on Earth. It follows that the specific impulse is measured in seconds. If it seems odd that this quantity is measured in units of time, it may help to know that the specific impulse can also be defined as the time in which 1 kg of propellant must be consumed to produce $1g$ of thrust. After all:

$$T = u \times \frac{\Delta m}{\Delta t} \quad \text{so that} \quad \Delta t = u \times \frac{\Delta m}{T} = u \times \frac{\Delta m}{\Delta mg} = \frac{u}{g} = I_{sp}$$

Table 3.1 lists the specific impulses for several propellants.

In practice, the choice of propellant is not simply a matter of picking a combination with the largest specific impulse. The difficulty in storing the material and the context in which it is to be used must be considered as well. For example, hydrogen and oxygen can remain in liquid form only when the temperature remains near absolute zero, which presents storage problems.

The Saturn V used RP-1/LOX in the first stage and LH_2/LOX in the second and third stages. Although the LH_2/LOX combination has a higher specific impulse, it was not used in the first stage as designing a high-efficiency multiple stage rocket demands that the first stage carries the bulk of the propellant mass for the whole rocket (see later in this chapter). As hydrogen's density is only 1/12 of the density of RP-1, the equivalent mass requires 12 times the volume. Storing that much LH_2 and LOX and keeping them ultra-cold to prevent their boiling into a useless gas was too difficult. The higher I_{sp} combination was used in the later stages as the mass of propellant required was much less.

In some situations, for example a satellite changing orbit or the service and lunar module engines used in Apollo, a very reliable mixture is required. The lunar module ascent stage had to fire first time or the explorers would have been stranded on the Moon. In such key circumstances, it is best to design an engine with the simplest mode of operation and the fewest number of parts that can go wrong. *Hypergolic propellant mixtures* are ideal for this as they spontaneously ignite when mixed, unlike the RP-1/LOX combination, for example, that requires an ignition source. All a hypergolic engine has to do is pump the chemicals from their storage tanks into a thrust chamber and the engine starts automatically.

A particularly simple form of rocket motor uses solid propellant. This is a mixture of both fuel and oxidiser in a semi-solid, putty-like form. Typically, this putty is packed around the inside of a cylindrical engine body with a hole left in the middle (Figure 3.1).

An igniter placed at the top sends a burst of flame into the hole. This lights the putty mixture at every point along the length of the engine. Thrust is produced when the hot gases push against the top of the cylinder. The exhaust gases shoot downward and out at the bottom. The level of thrust depends on how

Table 3.1 Specific impulses for a variety of propellant mixtures

FUEL	OXIDISER	I_{SP}/SEC	COMMENTS
Rocket propellant-1 (RP-1), a refined form of kerosene	Liquid oxygen (LOX)	303	LOX must be stored at cryogenic temperatures
Liquid hydrogen (LH_2)	LOX	453	LOX and LH_2 must be stored at cryogenic temperatures
Aerozine 50 (similar to kerosene)	Nitrogen tetroxide	320	Easily stored, comparatively cheap
Hydrazine	Nitrogen tetroxide	300	Corrosive, hypergolic
Aluminium polymer	Ammonium perchlorate	266	Solid propellant

Figure 3.1 Assembly of sections of a space shuttle solid rocket booster. The solid propellant lining the interior surface can be seen as the dark grey central ring. (Image courtesy of NASA and Kim Shiflett.)

much propellant burns at any one time. If the hole down the centre has a simple circular cross-section, increasingly large surface areas of new putty are exposed as the outer layers burn away. In such a simple design the thrust increases over time. Rocket engineers can produce very cleverly shaped holes in the putty, allowing some control of the variation of thrust during the flight.

The earliest rocket designs used solid propellants. However, as a solid rocket's storage tank and combustion chamber are essentially the same, its body must be very strong and so, heavy.[2] The first liquid-fuelled rocket (using gasoline and liquid oxygen) was launched by Robert Goddard on March 16, 1926. Liquid rocket development was important for practical rocketry, partly as the propellants have a greater specific impulse, but also as the engines can be modulated (throttling), stopped and restarted. Once you start a solid rocket it will continue until all the propellant is consumed.

Solid boosters were used as additional engines for the space shuttle during liftoff. Once their propellant was exhausted, they were jettisoned and parachuted into the sea (Figure 3.2) where they were recovered, refilled with propellant and used again on later launches.

3.4 APPLYING NEWTON'S LAWS TO SPACECRAFT

In order to produce the thrust required to lift the Apollo spacecraft off the ground, the engines of the Saturn V first stage consumed propellant at an average rate of nearly 12 tonnes per second. Once the first stage was exhausted, some 170 s into the flight, the rocket moved at a speed of 9,600 km per hour.

At first glance, these figures do not add up. From stationary to 9,600 km per hour in 170 s is an average acceleration of 15.7 m/s² – far greater than the acceleration quoted in the previous chapter (1.92 m/s²). However, a rocket's mass decreases as it burns. After all, propellant is consumed and ejected from the engines as exhaust. Even if the engines apply a constant force, the acceleration of the rocket will increase as its mass decreases. The acceleration calculated earlier was for a fully fuelled rocket at the moment of liftoff, not the average acceleration during launch. For that, we need a more sophisticated calculation.

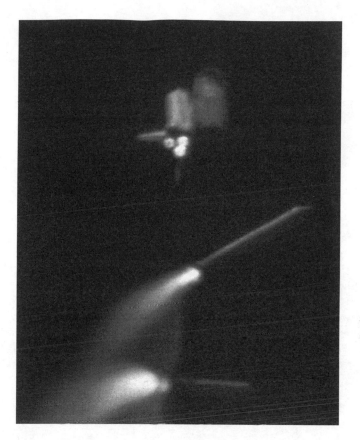

Figure 3.2 Solid rocket boosters separating from a space shuttle orbiter during launch. (Image courtesy of NASA.)

Newton's second law, as quoted in the previous chapter, (Force = Mass × Acceleration) works only if the mass of the accelerating object does not alter. To deal with a rocket changing mass during flight, the more developed form mentioned earlier in this chapter has to be used:

$$\text{Force applied} = \frac{\text{Change in momentum}}{\text{Time}}$$

which reduces to the earlier version in the special case of constant mass. After all:

$$\text{Momentum} = \text{Mass} \times \text{Velocity}$$

so

$$\text{Force applied} = \frac{\text{Change in (Mass} \times \text{Velocity)}}{\text{Time}} = \text{Mass} \times \frac{\text{Change in velocity}}{\text{Time}}$$
$$= \text{Mass} \times \text{Acceleration}$$

When the mass of the object changes, the relationship becomes:

$$\text{Force applied} = \frac{\text{Change in (Mass} \times \text{Velocity)}}{\text{Time}}$$
$$= \text{Mass} \times \frac{\text{Change in velocity}}{\text{Time}} + \text{Velocity} \times \frac{\text{Change in mass}}{\text{Time}}$$

or, using the standard abbreviations:

$$\text{Force applied} = \frac{\text{Change in (Mass} \times \text{Velocity)}}{\text{Time}} = \frac{\Delta(mv)}{\Delta t} = m\frac{\Delta v}{\Delta t} + v\frac{\Delta m}{\Delta t}$$

Whichever way it is presented, this expression is difficult to use because the various terms are not always constant; they change with time and hence we really have:

$$\text{Force applied} = m(t)\frac{\Delta v(t)}{\Delta t} + v(t)\frac{\Delta m(t)}{\Delta t}$$

where $m(t)$ means that the mass changes with time and $v(t)$ indicates the velocity changes.

To make further progress, we need specific expressions for $m(t)$ and $v(t)$. Newton had to invent a new branch of mathematics to deal with such situations: *calculus*.

For our purposes, the important matter is the final expression that can be derived for a rocket in flight. Two possible situations must be considered: (1) launch from the Earth's surface when gravitational pull needs to be taken into account and (2) manoeuvring the rocket in deep space when the effects of gravity can be ignored.

3.4.1 NO GRAVITATIONAL FORCE

The change in speed ΔV (pronounced 'delta V') that a rocket of mass M_R (the mass of the rocket's casing[3] empty of fuel plus the mass of the payload) can achieve by burning propellant of mass M_P in the absence of a gravitational force is:

$$\Delta V = u \times Ln\left(1 + \frac{M_P}{M_R}\right)$$

where u represents the speed at which the exhaust gases are ejected from the back of the rocket. Ln is the natural logarithm. The ΔV expression is known as the *Tsiolkovsky rocket equation*, after Konstantin Tsiolkovsky, a pioneer Russian rocket scientist who derived it independently in 1903. The equation was derived earlier by William Moore, a British mathematician, in 1813 and later by William Leitch in 1861. Tsiolkovsky is credited with the equation because he was the first to apply it to the question of rockets achieving the speeds necessary for space flight. If you are interested in seeing how the equation is obtained, the proof is set out in Appendix 2.

We can rewrite the equation above to include the specific impulse (I_{sp}):

$$\Delta V = g \times I_{sp} \times Ln\left(1 + \frac{M_P}{M_R}\right)$$

Sometimes this version of the equation causes confusion due to the presence of g (strength of Earth's gravity) even though the equation is applied in deep space. This term enters the formula via the definition of specific impulse and does not imply the presence of gravity.

ΔV is the currency of rocketry. Any manoeuvre in space requires a velocity change (ΔV) and this equation establishes the amount of propellant needed. Given that spacecraft carry limited amounts of propellant, every alteration of orbit, docking manoeuvre or speed change must be accounted for carefully. The ΔV budget of a complete mission must be calculated in advance, with some latitude for emergencies. Every kilogram of propellant carried on a spacecraft constitutes mass that could have been used in the payload. Mission designers are very thorough in trying to eliminate excess mass.

It is worth noting that the time taken to achieve a specific ΔV is not part of the expression: the length of the burn does not affect the outcome, except via the mass consumed. This is true only if no forces other than engine thrust act on the rocket. In deep space, where gravity can be neglected, this equation would

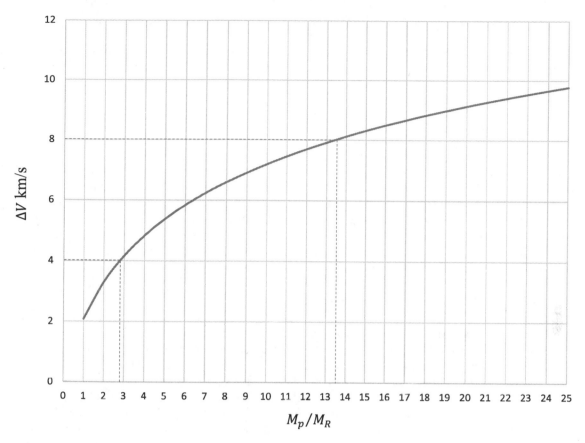

Figure 3.3 How the ΔV achieved by a burn varies with mass ratio.

apply. In that case, if the propellant burns slowly, the rocket will take a long time to achieve its final speed. On the other hand, if the propellant is burnt quickly, that speed will be reached in a far shorter time, but the final speed will be the same in both cases.

The ratio between the mass of propellant burnt and the mass of the rocket (casing + payload) is critical. To achieve high speed, the mass of propellant burnt must be much greater than the mass of the rocket.

Figure 3.3 shows how the ΔV of a hypothetical rocket with an exhaust speed of 3 km/s varies with burn mass ratio. Assuming an initial speed of zero, the ΔV is the same as the final speed. The curve establishes the severe penalty that must be paid for increasing the ΔV. Doubling the ΔV from 4 to 8 km/s requires an increase in mass ratio from just under 3 to more than 13. Assuming the mass of the empty rocket remains the same, the mass of fuel must increase by a factor of 13/3 or 4.3. Of course, it is not always possible to increase the mass of propellant available without increasing the mass of the empty rocket, since the storage tanks will have to be larger and heavier. In any case, even more propellant would have to be carried in order to slow the rocket at the end of the journey (often any ΔV that you put in has to be taken out again…).

Ratios very much greater than 10:1 are impractical since the tremendous weight of the propellant has to be supported by robust tanks that add to the weight of the rocket and hence reduce the ratio. Exhaust speeds greater than 3 km/s (about nine times the typical speed of sound at sea level) are also difficult to achieve. The Saturn V's first stage produced an exhaust velocity of 2.7 km/s and had an excellent mass ratio of 15:1.

3.4.2 THE EFFECT OF GRAVITATIONAL PULL: LAUNCH

The version of the ΔV equation considered so far is valid only when no gravitational or other forces are acting on a spacecraft. While this may be a reasonable approximation in deep space, it is certainly not true during launch, and was not true during the flight of Apollo which was subject to gravity from both the Earth and the Moon.

Fortunately, the required modification is minor, although the effect is quite important. In the presence of a gravitational field of strength g, the ΔV achieved by burning propellant of mass M_P is:

$$\Delta V = g_E \times I_{sp} \times Ln\left(1 + \frac{M_P}{M_R}\right) - gt$$

Note that I have used g_E to indicate the strength of Earth's gravity.

The most significant aspect of this new expression is that time has been introduced. In the presence of gravity, the ΔV is no longer independent of burn time. The longer it takes to consume the propellant, the smaller the ΔV. This is why it is vital to pump propellant into the engines at massive rates during launch.

One way of making this more intuitively obvious is to remember that the thrust achieved by the engines must be greater than the weight of the rocket, or it will never leave the ground. The greater the thrust, the faster the rocket will accelerate – in other words, achieve a greater ΔV in a shorter time. In order to increase the thrust, propellant must be used at a rapid rate and we are back to the same conclusion again.

3.5 REAL ROCKET ENGINES

Translating basic rocket motor physics into a useable engine is a complex engineering task. Many components must be optimised to ensure that engines deliver the maximum possible thrust from the propellant. The Saturn V used two engine designs.

The F-1 (Figure 3.4), running on RP-1/LOX, powered the first stage. The other stages used the J-2 design running on LH$_2$/LOX. Five of the F-1 engines were used on the first stage, the outer four mounted on gimbals to allow up to six degrees of tilt, giving some measure of steering ability over the rocket's trajectory. Figure 3.5 illustrates some common features of all rocket engine designs. It is not intended to be a faithful rendition of the F-1 or J-2 engines.

3.5.1 PROPELLANT DELIVERY

One of the primary goals of good engine design is the delivery of fuel and oxidiser to the thrust chamber in the correct proportions at a correct rate. In the F-1, fuel was delivered at 58,000 litres per minute and oxidiser at 93,900 litres per minute. Both liquids fed into the thrust chamber via pumps driven by a common shaft from a gas turbine. The high-velocity gas required to drive this turbine was derived from a gas generator in which small amounts of fuel and oxidiser were mixed. At the moment of engine ignition, valves opened to allow pressurised fuel and oxidiser into the gas generator where they were ignited by small explosives. The reaction produced exhaust gases that passed through the turbine blades at 77 kg/s. The turbine, generating 41 MW, drove the pumps which channelled more propellant to the gas generator and to the thrust chamber.

It was important to maintain constant pressure in both the fuel and oxidiser tanks during the flight. If a partial vacuum was allowed to develop above the liquids (as it does whenever liquid is drained from a sealed container), the pumps would have a harder time drawing liquid from the tanks. As the levels dropped, evaporated liquid would partially fill the empty space, but there would still be a pressure loss in the engine feed lines. To offset this, helium was pumped into the RP-1 tank and gaseous oxygen derived from the LOX tank was used to maintain that tank's pressure.

A similar problem would develop in flight at the moment of staging. Once a stage's engines shut down and before the next stage lit up, a rocket would fall freely under gravity (although, of course, it would continue to move upward). Under these circumstances, propellants tend to float in the tanks and become difficult to pump. To counter this, small ullage motors would fire in the next stage prior to the main motor ignition. This would give the stage an upward acceleration and settle the liquids before the pumps cut in.

Both the thrust chamber and the engine bell had to be cooled during operation to prevent them from melting under the extreme temperatures of the exhaust gases (3,300°C in the F-1 engine thrust chamber).

Figure 3.4 Wernher von Braun standing next to a Saturn V first stage with its five F-1 engines. (Image courtesy of NASA.)

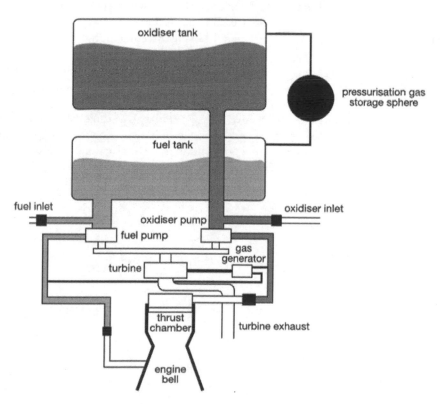

Figure 3.5 A schematic diagram illustrating basic engine design. The fuel reaches the thrust chamber by circulating in pipes around the outside where it acts as a coolant.

Rocketry

The F-1 thrust chamber was kept cool by circulating cryogenic fuel around the outside before the fuel entered to react with the oxidiser. The engine bell was cooled by routing turbine exhaust around the structure.

3.5.2 NOZZLE DESIGN

At the far end of the thrust chamber, the exhaust gases emerge into the surroundings via a *nozzle*, the purpose of which is to convert as much of the random motion (pressure) in the exhaust gas into concerted motion (flow) as possible.

The balloon mentioned in Chapter 2 does not provide efficient thrust because there is no control over the direction in which the molecules emerge from the hole in the skin. If the best use is to be made of the energy in the molecules, they should all emerge parallel to the direction of flight. Remember that the thrust results when molecules strike the skin of the balloon on the opposite side (in a real engine, the exhaust molecules strike the top of the thrust chamber). The random nature of molecular motion in both cases ensures that the molecules providing the thrust strike the balloon skin (thrust chamber) in all directions. In many cases, the forces cancel each other out.

In Figure 3.6, molecules A and B will exert forces on the balloon's skin, but would not fully propel the balloon forward. This is because they strike the balloon at an angle to the direction of flight. Part of the impact force of molecule A will push the balloon upward and part of the impact force of molecule B will push the balloon downward. Not only do the pushes occur at an angle to the direction of flight, but the impacts of both molecules are partially offset against each other. Molecules C and D carry some energy as they leave the balloon. They ensure that the impacts of A and B provide net thrust to the balloon, but not as efficiently as the horizontal pair of molecules shown in the same diagram. If the hole in the balloon could be designed so that only molecules moving parallel to the direction of flight left the balloon, the thrust produced would be more effectively employed. One way to do this is to reduce the size of the hole. A small hole will allow molecules out only if they approach almost straight on. However, a natural consequence is that fewer molecules can get through. As the thrust is partially dependant on the rate at which mass is consumed (or ejected from the balloon), reducing the size of the hole also reduces the thrust.

A real engine must be designed more cleverly to ensure that as much mass as possible is allowed out of the engine per second and that the molecules exit parallel to the direction of flight and move as rapidly as the design permits.

A gas exerts *pressure* because its molecules move about rapidly in random directions. As a result, they collide with each other and with the walls of their container. These collisions exert forces on the walls that translate into pressure.

Gas will *flow* when its molecules have a concerted drift speed in a given direction, in addition to their rapid random motions. Consider the atmosphere in the Apollo command module. The gases inside the capsule will exert pressure due to their random motion. However, the molecules must also participate in the forward motion of the capsule; otherwise they would be left behind as the craft travels along.

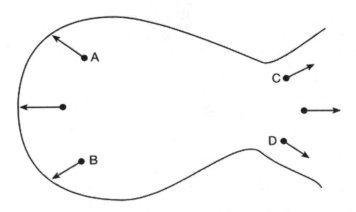

Figure 3.6 Thrust in a balloon.

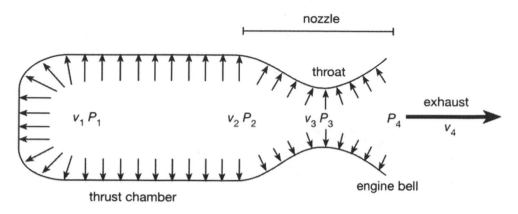

Figure 3.7 A convergent–divergent nozzle design for a rocket engine. The arrows represent the pressure exerted by the gas. Left: Very low flow velocity of the exhaust gas, v_1. Right: v_4 is hypersonic.

At the top end of the thrust chamber, the gas exerts high pressure but does not flow fast. At the bottom of the engine bell, the gas flows quickly but its pressure is less. In fact, the pressure is equal to the external pressure in the air.

The nozzle acts to convert the gas pressure at the throat (Figure 3.7) to a directed flow of exhaust from the engine bell. This has two effects: (1) if the consumed propellant leaves the thrust chamber as exhaust at the greatest mass rate per second, more propellant can be pumped in to continue the reaction and (2) the conversion of pressure into high-speed flow causes a back reaction that maximises the pressure at the top of the thrust chamber. This pressure pushes the rocket forward.

The most common design is the *de Laval nozzle*, named after Dr Carl de Laval, a Swedish engineer who first conceived the idea. The nozzle consists of a gently tapering exit from the thrust chamber down to a minimum diameter (the throat) and an expanding area (the engine bell), opening onto the surroundings (Figure 3.7).

The tapering exit hole leading to the throat increases the flow of gas leaving the thrust chamber in the same way that putting your finger over the end of a hosepipe will increase the speed at which the water escapes. In fact, two things are happening. First, the tapering exit hole preferentially selects molecules that are travelling parallel to the centre line of the engine. Second, molecules that bounce off the sides are not escaping. They collide with molecules arriving from the top of the chamber. This has the effect of building the density of gas behind the throat. Consequently, the pressure, P_2, increases there (and back through the chamber, i.e. P_1). As a result of this pressure buildup, the number of collisions acting on molecules by pushing them *down* the throat exceeds the number of collisions that push *back*, resulting in a net acceleration of the molecules into the throat of the nozzle. The amount of acceleration the molecules receive depends on the pressure difference between the start of the nozzle and the throat (P_2 and P_3 on the diagram).

With more molecules travelling parallel to the sides of the throat, fewer of them strike the sides and so the pressure of the gas on the side walls is reduced. This effect is commonly used in various engineering applications.[4]

The taper on the throat of a rocket engine is designed to increase the rate of gas flow up to the speed at which sound waves will travel through the exhaust gas. Technically, this condition is known as a *choked nozzle*.[5] Of course, this does not mean that the throat is choked physically; exhaust gases driven by the pressure difference can still travel down the throat.

The speed at which a sound wave travels through a gas depends on the motions of the gas molecules. A sound wave is a sequence of *compressions* (high-pressure regions) and *rarefactions* (low-pressure regions) that move through the gas in a pattern (Figure 3.8).

In a gas which is not flowing, a disturbance travels due to collisions between the molecules. The difference between a sound wave and a draft of wind is that the molecules do not move from their average positions in a sound wave; a draft of wind involves a net drift of molecules. This drift speed is superimposed on the molecules' random thermal motions. A sound wave cannot travel through a gas at a speed greater

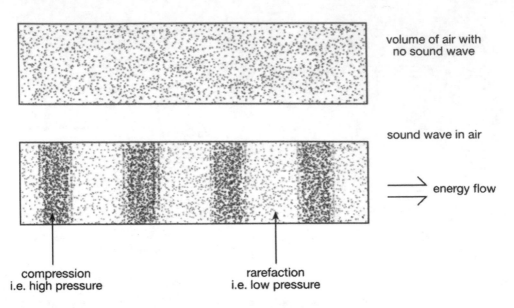

volume of air with
no sound wave

sound wave in air

energy flow

compression
i.e. high pressure

rarefaction
i.e. low pressure

Figure 3.8 Top: A pipe containing air. The dots represent molecules in rapid random (thermal) motion. The molecules are travelling at great speed, but usually go nowhere. Bottom: A pipe with a sound wave passing through it. The molecules cluster into high-pressure and low-pressure regions. A moment later, the high-pressure regions will have become low-pressure regions and vice versa. This reversal arises from collisions between the molecules, hence sound cannot travel at a speed that exceeds the thermal motion of the molecules.

than the average speed of the molecules *due to their thermal motion*. However, there is no reason why *drift speed* cannot exceed *thermal speed*. When this happens, the gas flows faster than the speed of sound; this is called *hypersonic flow*.

A typical engine design will arrange for the ratio between P_2 and P_3 to be great enough to *apparently* accelerate the gas in the throat to hypersonic speeds. In fact, in the situation of a constricted throat, the gas can be accelerated only up to the speed of sound. Instead of exceeding this speed, the pressure difference compresses the gas. This is how things stand as the exhaust moves beyond the throat into the engine bell.

The gas flowing through the throat is no longer sensitive to pressure changes that occur beyond the end of the nozzle.[6] A sound wave is a pressure wave communicated by thermal motion and if molecules drift forward at speeds greater than their thermal motion, there is no way for any pressure change to communicate back through the gas. When the ratio of P_2 to P_3 (Figure 3.7) is great enough for the gas to be accelerated to the speed of sound, any changes in P_4 will have no effect on the speed in the throat.

Under normal circumstances when a gas arrives at a widening aperture, such as an engine bell, the speed of the molecules will start to drop. However, a compressed sonic velocity gas generates a different and complex effect. The compressed gas expands explosively, releasing energy into the speed of the molecules.[7] The result is that the molecules accelerate and exit the engine bell at hypersonic velocity.

The exact shape of the engine bell has to be very carefully designed. The trick to obtaining the greatest efficiency of thrust is to match the pressure of the gas emerging from the end of the bell to the pressure of the surrounding atmosphere. Remember that the sideways pressure of the gas is due to imperfect alignment of molecules along the direction of flight. If the external atmospheric pressure exceeds the sideways pressure of the exhaust gas as it leaves the engine bell, collisions between the molecules in the atmosphere and those in the exhaust will push the exhaust inward. This decreases the efficiency of the thrust. On the other hand, if the exhaust gas pressure exceeds that of the atmosphere, the net tendency is for molecules in the exhaust to move outward, again reducing efficiency (Figure 3.9).

Clearly, as a rocket travels higher, the pressure of the atmosphere decreases (it is zero in space) so that engine bells cannot be optimised for all altitudes. The F-1 and J-2 engines had nozzles of different shapes because the stages in which they were installed were designed to work at different altitudes.

Rocketry

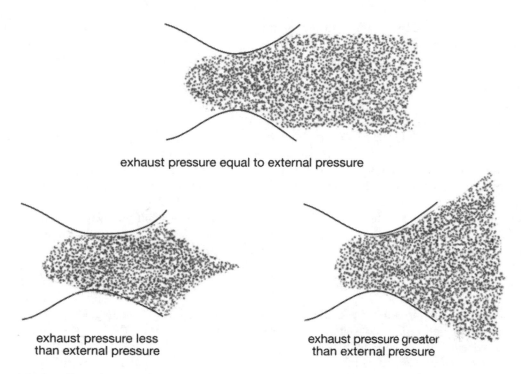

exhaust pressure equal to external pressure

exhaust pressure less
than external pressure

exhaust pressure greater
than external pressure

Figure 3.9 The effect of external pressure on the shape of an exhaust plume from a rocket engine.

There is another way of looking at this argument. Figure 3.7 shows only the pressures acting within the system of the thrust chamber and nozzle. In the atmosphere, the pressure of the air acts externally and equally over the entire shape of the rocket. Air molecules striking the front of the ship tend to slow it down.[8] Air molecules striking exhaust molecules exiting the engine bell will tend to bounce them back. These molecules then strike other molecules leaving the bell, pushing most of them back. This effect provides a forward push on the rocket. Consequently, the total force acting on a rocket in flight in the atmosphere is:

$$F = u \times \frac{\Delta m}{\Delta t} + (P_4 - P_A) \times A_B$$

where P_A denotes the atmospheric pressure, P_4 represents the exhaust gas pressure and A_B is the area of the exit hole of the engine bell.

The first term in this equation is the *momentum thrust* covered earlier. The second is the *pressure thrust* arising from the pressure difference between the front and rear of the rocket as discussed above. At first glance, this equation implies that the greatest force on the rocket would develop if the exhaust pressure was much greater than atmospheric pressure. However, what is not obvious from the bare terms in this equation is that u (exhaust speed) *also* depends on exit pressure. A large exhaust gas pressure implies little flow of the gas. The molecules emerging from the bell have a large random motion (pressure). As a result, the overall velocity of the gas from the bell (u) is comparatively small (the molecules move quickly in random directions but do not move very far). A large P_4 implies that u is small.

Rockets are designed to consume fuel at a fast rate and hence the $\Delta m/\Delta t$ value is very large. Consequently, momentum thrust is a much greater contributor to overall thrust than pressure thrust. Surprisingly, an increase of P_4 will *reduce* the overall thrust, as the smaller value of u implied will have a much larger effect. Conversely, if the exit pressure is less than atmospheric pressure, the pressure behind the rocket is less than the atmospheric pressure acting backward on its nose. Furthermore, the exhaust gases struggle to exit the engine bell in a smooth and linear flow and this combination also produces less thrust. The greatest overall thrust occurs when P_4 equals P_A and u is hence a maximum. For this reason, engine bells are designed to bring the exiting exhaust gas up to a pressure that equals the external pressure making them most efficient at a specific altitude.

3.5.3 CONTROLLING THRUST (THROTTLING)

From the earliest stages of the lunar mission design, it was clear that one of the greatest technical challenges was the lunar module (LM) descent engine. The demands of a powered descent to the surface, coupled with the need for a spacecraft to fly around while searching for a safe landing site, mandated an engine with variable power output: the engine had to be *throttleable*. This was a revolutionary requirement for a rocket engine.

During the development of the LM, NASA instructed Grumman, its main contractor, to subcontract to two companies for the preliminary design of the descent engine. Rocketdyne's engineers, who were responsible for the F-1 and J-2 engines already in use in the Saturn V, suggested that a throttleable engine could be produced by introducing an inert gas such as helium into the propellant to alter the mixture without changing the flow rate. This was a new approach to engine design that would not require adapting pumps to deal with variable pressure.

A rival idea was put forward by Space Technology Laboratories (STL). They suggested a pressure-fed hypergolic system with variable flow rate valves to provide throttle adjustment. The design also called for an injector system that had a variable area (a concept similar to that used in some shower head designs). Grumman eventually recommended that the Rocketdyne design be approved, but was overruled by NASA[9] and the contract went to STL.

3.6 STAGING

One of the often-quoted virtues associated with an active space program is the pressure exerted on the development of new technologies. For the Moon missions, some materials had to be designed to survive the heat of re-entering Earth's atmosphere. Other materials, used to construct rockets, had to meet very strict weight requirements. For the Saturn V, Alcoa and Reynolds collaborated to develop an extremely strong aluminium alloy that could be produced in sheets up to 0.64 cm thick for the external skin of the rocket. Given the use of such thin, lightweight material the design had to be carefully contrived to support the tremendous weight of propellant. A fully fuelled Saturn V was some 20 cm shorter than it was when empty due to the weight of stored liquid onboard.

Once the propellant has been chosen, the specific impulse is fixed and high mass ratios become the goals of a design. Even then, as noted, it is difficult to achieve a ratio between the masses of propellant and rocket that is much higher than 10:1. To make matters worse, an engineer simply cannot stuff a rocket full of greater and greater amounts of fuel in the hope of achieving higher final velocity. The physical reason is simple. The more fuel in the rocket, the heavier it is and the harder the engines must work to lift it off the ground.

Figure 3.10 makes the point graphically. N is defined as the ratio between the total mass of the rocket (i.e. casing + fuel + payload) and the mass of the payload. Along the curve, the ratio R between the mass of the empty casing and the fully fuelled casing (not counting the payload) has been fixed at 0.1. This rocket casing is ten times heavier when fuelled. The exhaust speed is constant at 3 km/s. The curve illustrates what would happen if we were to launch the same mass of payload by using bigger and bigger rockets.

Even with an impracticably large rocket 700 times the mass of the payload, the final velocity is not much more than that achieved by a rocket only 50 times the mass of the payload. A fully fuelled Saturn V sitting on a launch pad had a total mass some 56 times that of the Apollo spacecraft.

Clearly, no matter how heavy this rocket is, it will never reach a final speed of 7 km/s. Single stage rockets are fundamentally limited by the fuel: mass ratio (R) that can be achieved.

Rocketry

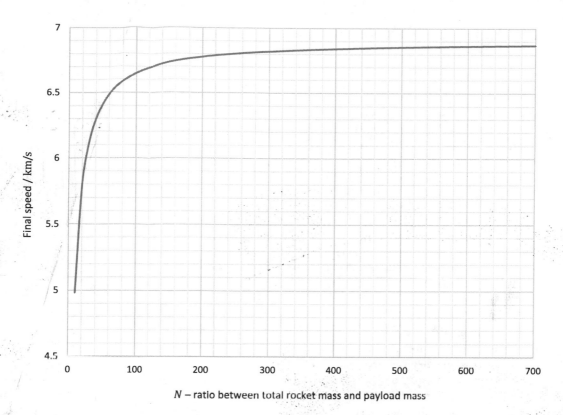

Figure 3.10 The final speed for a single stage rocket with an exhaust speed of 3.0 km/s.

Single stage rockets

The limitations of a single stage rocket can be demonstrated by 'fiddling' with the ΔV equation. The easiest way is to work in terms of multiples of the payload mass, as in Figure 3.10. For the sake of simplicity, I will use the equation that does *not* include the effect of gravity, but the conclusion will be the same. I assume that the payload carries no propellant (or that the propellant it contains is not used as part of the launch sequence, as was the case with Apollo) and that it has a mass m. The casing mass (the empty rocket without payload or propellant but including propellant tanks and engines) is M_C and the propellant mass is M_P.

The total mass of the whole rocket, including payload, engines, propellant and casing, is N times the payload mass or Nm. Finally, I will set R to be the ratio between the mass of the empty casing and the mass of the fuelled casing. We assume that $R = 0.1$ for most practical rockets.

Summarising

Mass of payload $= m$
Mass of whole rocket at launch $= Nm = m + M_C + M_P$
Mass of fuelled casing $= Nm - m = m(N - 1)$
Mass of casing $= R \times$ mass of fuelled casing $= Rm(N - 1)$
Mass of rocket after fuel is used up $= M_R = M_C + m = Rm(N - 1) + m$
Exhaust velocity $= u$

Inserting all these values into the ΔV equation gives:

$$\Delta V = u Ln\left(1 + \frac{M_P}{M_R}\right) = u Ln\left(\frac{M_P + M_R}{M_R}\right)$$

$$= u Ln\left(\frac{Nm}{Rm(N - 1) + m}\right)$$

When the dust settles, this boils down to:

$$\Delta V = u\,Ln\left(\frac{N}{R(N-1)+1}\right)$$

which is the equation used to plot the graph in Figure 3.10.

We can simplify this expression further, as once N becomes as large as 1,000 or more, the difference between N and $N - 1$ can be ignored. Similarly, if RN becomes at least 100 then $RN + 1$ is practically the same as RN. We now have:

$$\Delta V = u\,Ln\left(\frac{N}{R(N-1)+1}\right) \approx I_{sp}g\,Ln\left(\frac{N}{RN}\right) = I_{sp}g\,Ln\left(\frac{1}{R}\right)$$

The final result shows that the ΔV achieved with an enormous rocket *does not depend on the mass of the rocket* and is *limited by the fuel: mass ratio*. In this instance, with an exhaust velocity of 3.0 km/s and R of 0.1, the maximum ΔV is 6.9 km/s, exactly as seen in Figure 3.10. Interestingly, this is less than the speed required to escape from Earth's gravity.

The answer is to construct a rocket from multiple stages, each of which has its own fuel tanks and motors. At first glance, this appears to defeat the object as adding the mass of the engines needed for each stage will reduce the mass ratio of the rocket. However, the ability to discard the empty stage with its engines and the empty fuel tanks that are no longer required – a vital weight reduction – more than compensates for the extra initial mass. As the second stage builds its ΔV on top of that already achieved by the first stage, an overall performance improvement can be gained. The process is a little like throwing a ball from a moving train. The ball ends up with considerable speed as it already has the speed at which the train is moving. If each stage has a mass ratio of 10:1, the performance is better than that of a single stage booster with the same ratio.

The first stage of a multiple stage rocket has the most powerful engines and carries the largest mass of fuel by far (see the maths box 'Multiple stage rockets' below). First stages are generally designed to carry a rocket into the upper atmosphere where the air density is lower and so the air resistance is far less. The second stage then fires and carries the payload almost into orbit. With the Saturn V, a burn from the third stage was required to accelerate the Apollo up to the required speed for Earth orbit.

Multiple stage rockets
To see the advantages of multiple stage rockets in more quantitative terms, consider a two-stage rocket where each mass is quoted as a multiple of the payload (Figure 3.11).

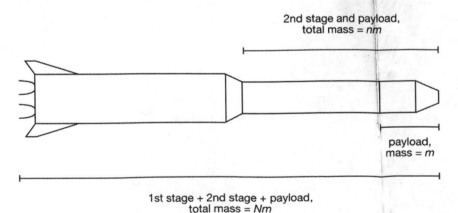

2nd stage and payload, total mass = nm

payload, mass = m

1st stage + 2nd stage + payload, total mass = Nm

Figure 3.11 Parameters for the analysis of a multiple stage rocket.

Once again R is the ratio of the mass of the empty casing to the mass of the fuelled casing.

Mass of first stage with fuel $= Nm - nm$

Mass of first stage without fuel $= R(Nm - nm)$

Therefore, after the first stage has burnt out, the mass of the unseparated rocket as here:

$$M_R = nm + R\,(Nm - nm) = nm\,(1 - R) + RNm$$

$$\Delta V_1 = uLn\left(1 + \frac{M_P}{M_R}\right) = uLn\left(\frac{M_P + M_R}{M_R}\right)$$

$$= uLn\left(\frac{Nm}{nm(1 - R) + RNm}\right) = uLn\left(\frac{N}{n(1 - R) + RN}\right)$$

which does not depend on the payload mass at all.

A similar calculation produces the ΔV achieved by the second stage burn:

$$\Delta V_2 = uLn\left(\frac{n}{R(n - 1) + 1}\right)$$

Now, the total ΔV after the two stages have burnt out will be the sum of the ΔVs achieved by each stage in turn:

$$\Delta V = \Delta V_1 + \Delta V_2$$

which depends on the relative sizes of N and n.

Figure 3.12 shows how the total ΔV for a two-stage rocket varies with the mass of the second stage. The total mass of the rocket has been fixed as Nm with $N = 50$. Consequently, the mass of the second stage (nm) must be such that n is less than 50. The graph clearly shows a peak at a low value of n. In other words, the largest overall speed is achieved if most of the mass is in the first stage. This is partly why RP-1 was chosen as the fuel for the Saturn V first stage: it is much denser than LH_2 and so a high fuel *mass* could be achieved in the first stage without needing a large *volume* to store fuel. A detailed calculation, using calculus shows that the optimum is achieved if n equals the square root of N (Appendix 2).

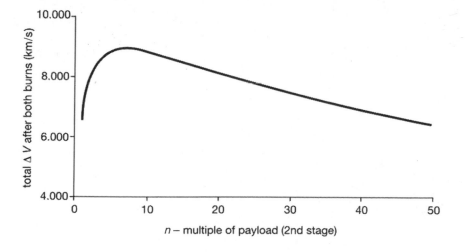

Figure 3.12 Total ΔV achievable as a function of n.

For a rocket with an overall mass equal to 50 times the payload mass, the greatest ΔV comes about if the first stage has a mass of $\sqrt{50}$, or 7.07, which is where the peak appears on the graph.

Interestingly, fixing $n = \sqrt{N}$ also has the effect of making the ΔVs achieved by each stage the same. However, we must remember that this conclusion does not take into account the role of gravity and the enormous air resistance that acts on the first stage in particular (as the rocket flies through the thicker atmosphere). In the specific case of the Saturn V, the three stages performed as per Table 3.2. Comparing the mass of propellant with the casing mass for each stage drives home the point about the large mass ratios used.

Table 3.2 Performance data for Saturn V stages

STAGE	BURN TIME (SEC)	PROPELLANT MASS (TONNES)	CASING MASS (TONNES)	BURNOUT VELOCITY (KPH)	BURNOUT ALTITUDE (KM)
1	170	2,147	136	9,654	61.1
2	395	444	40	24,617	184.2
3	165 into orbit	107	13	24,617	185.0

3.7 FUTURE DEVELOPMENTS IN ROCKETRY

Looking to the future, it is possible that chemical rockets will be abandoned for many applications except launches from Earth's surface and short-range hauls (as far as the Moon). This is largely as a result of the limit on specific impulse obtainable by chemical means. The best propellants have specific impulses ~450 s, so in order to achieve the ΔVs needed to explore the solar system, large mass ratios must be employed. This is unfortunate as such missions will require sustaining astronauts in space for long periods (measured in months, not days as for Apollo), so the payloads will be far heavier (consumables such as oxygen, water and food constitute a substantial fraction of the mass). If the payload is heavier, achieving a given mass ratio means that the total spacecraft mass must be far heavier as well.

It is clearly advantageous to increase the specific impulse as much as possible, allowing the mass ratio to decrease so that more of the mass of a spacecraft can be devoted to the payload. This is illustrated by Figure 3.13 which shows how the mass ratio required to achieve a given ΔV varies with different specific impulses.

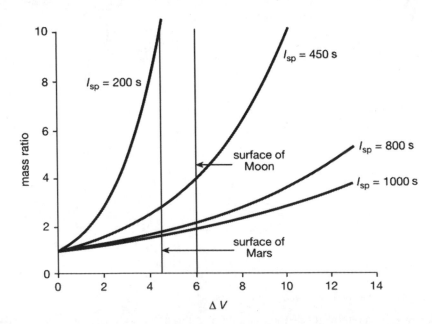

Figure 3.13 How the mass ratio required for a given ΔV varies with specific impulse. The reason why less ΔV is required to reach Mars than the Moon, is explained in Chapter 4.

Remember that in practical terms a mass ratio much greater than 10:1 ($R = 0.1$) cannot be achieved. There may be some hope for increasing this slightly, as lower density alloys and composite materials become available. However, this is not going to push mass ratios as high as 100:1. On the other hand, raising the specific impulse to ~1,000 s would have a dramatic effect on the mass ratio.

Increasing the specific impulse implies producing greater exhaust velocity. The flip side is that these designs do not consume propellant at a very fast rate; hence the thrust they produce is very small. However, remember that ΔV depends on exhaust velocity and mass ratio, not on the rate at which propellant is used. Such engines would be hopeless for lifting loads off the ground, but in space, they can reach enormous speeds by accelerating gently for most of the journey.[10]

3.7.1 NUCLEAR THERMAL ENGINES

A nuclear reactor is an efficient mechanism for generating heat. Fission reactors work by splitting the nuclei of very heavy elements into lighter elements, a process that releases energy. One common nuclear fuel is uranium. Some uranium nuclei, with a specific arrangement of protons and neutrons, will split when a slow-moving neutron collides with them:

$$\text{Neutron} + \text{Uranium nucleus} \xrightarrow{\text{split into}} X + Y + \text{Some number of neutrons}$$

X and Y are the nuclei of some lighter elements (it is impossible to predict what X and Y will be in any specific case; the splitting process is random). Also produced are two or three neutrons and some energy in the form of gamma rays and exotic particles called *neutrinos*.

The neutrons produced by the reaction can go on to collide with other uranium nuclei and cause further reactions that will, in turn, produce more neutrons and other particles. This continuing process is called a *chain reaction*. Once a chain reaction is started, by introducing neutrons into the reactor, the reaction becomes self-sustaining, if the system is engineered correctly. The neutrons produced by the reaction have to have a good chance of colliding with other uranium nuclei, for example. In a practical reactor, the reaction is regulated by *control rods* – cylinders of a neutron-absorbing material that are lowered into the reactor core. The further inside the core the control rods sit, the greater the number of neutrons that are absorbed and the slower the reaction rate, and vice versa.

Most of the energy produced in the reaction is the kinetic energy of the products X and Y, which are produced moving at considerable speeds. In a power station reactor, the uranium fuel is normally stacked in discs and contained in fuel rods. Consequently, X and Y cannot go very far without colliding with other nuclei inside the fuel rod. After several such collisions, their kinetic energy spreads throughout the rod, raising the temperature. Coolant circulating through the reactor extracts thermal energy and, in the case of an electricity generating station, the hot coolant is used to boil water into steam which drives the turbines that generate electricity.

A somewhat similar process can be used to produce thrust. A compact nuclear reactor, such as those found on nuclear submarines (which utilise a slightly different operating principle from that described above) can serve as a heat source for a spacecraft.

The thermal energy extracted from the reactor boils a propellant liquid (Figure 3.14) The combination of pressurised gas in the thrust chamber and escaping gas from the nozzle provides thrust. The idea is rather like using the steam produced by a kettle to propel the kettle across a room. On Earth, this would not work, but in space, a kettle would gently push itself along in a path opposite to the escaping steam.[11]

Hydrogen is a suitable propellant for a nuclear engine. As it is the lightest element, its molecules have the advantage of yielding the highest speeds at certain temperatures. The potential disadvantage of having to store the liquid at very low temperatures is more easily overcome in space than on the ground.

A nuclear engine would never produce sufficient thrust to lift a spacecraft off the ground. However, it would provide excellent propulsion for trips between planets, as the gentle thrust can be maintained for very long periods (weeks) so considerable speeds can be achieved.

In the 1960s, NASA carried out development work on nuclear engines. The Nuclear Engines for Rocket Vehicle Applications (NERVA) programme constructed and tested several variants with thrusts

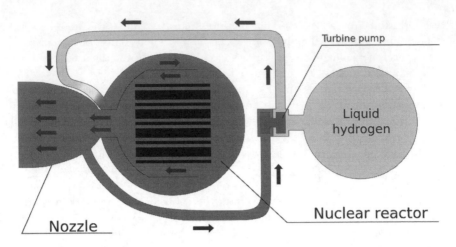

Turbine pump

Liquid hydrogen

Nuclear reactor

Nozzle

Figure 3.14 Basic design features of a nuclear thermal engine. (Image courtesy of Wikipedia Commons.)

between 44,000 and 1,111,000 N. These engines delivered specific impulses of the order of 800 s – much better than a chemical rocket. At the time, the tests were taking place with a manned Mars mission in mind. Consideration was given to using them in a modified Saturn V third stage. However, when the Nixon administration cut the NASA budget, NERVA was a casualty. Despite this, the principle of the nuclear engine was established successfully and over 17 hours of accumulated ground-based test firing experience was achieved. Indeed, one of the engine designs was certified as flight-safe before the programme came to an end and nuclear engines remain valid options for interplanetary exploration. The Russians have also tested nuclear thermal propulsion successfully.

Despite their evident virtues, nuclear engines have two drawbacks. First, it is difficult to ensure efficient heat exchange between the reactor coolant and the liquid hydrogen. A great deal of energy is wasted in the exchange process. The efficiency is related to $\sqrt{T}$ where T is the temperature of the core. While exceptionally high temperatures are achievable in principle, in practice the melting point and other thermal parameters of the materials involved tend to be the limiting factors.

Second, as noted above, nuclear engines could not be used to lift spacecraft off the ground. Therefore, at some point, a nuclear reactor or the components for constructing one would have to be launched by standard chemical means. Environmental considerations make this a contentious proposition. The consequences of scattering radioactive material across a wide area if a rocket should explode are potentially very serious.[12]

The Russians have launched about 40 reactors of various designs into space, but more common are *radioisotope thermoelectric generators* (RTGs), which use the heat generated by radioactive decay to produce electrical power. These are not reactors as such, but they are constructed from potentially hazardous materials. The Apollo missions routinely carried RTGs in their experimental packages and deep space probes use RTGs to provide electrical power where the sunlight is too weak to generate solar electricity. The active components of RTGs are heavily shielded to protect astronauts from radiation on manned missions and to provide robustness during launch.

3.7.2 SOLAR THERMAL ENGINES

The solar thermal engine employs the same basic concept as the nuclear engine: boiling a propellant liquid to provide thrust. However, the heat source in a solar engine is focussed sunlight, which is far more practical in space where there are no clouds to block the view of the Sun. This engine also has the advantage of not needing radioactive materials to be launched into space. However, to generate the temperatures required, concentrated sunlight is needed and this makes solar thermal engines less effective in the outer solar system. Also, projections indicate that the engine can develop only about 400 N of thrust.

3.7.3 ION DRIVE (ELECTRICAL PROPULSION)

The idea behind an ion engine is to accelerate ions (charged atoms) to very high speeds and fire them from the back of a spacecraft. Like the nuclear engines, an ion drive would be useless for lifting objects off the ground. However, they can be used to provide gentle thrust for considerable lengths of time.

Many variations of ion drive have been and are currently in development. In a Gridded Electrostatic Ion Thruster, the propellant is ionised by bombardment with electrons from a warm cathode (negatively charged terminal). These electrons are accelerated through the region where the gas enters the working chamber by attraction to the high voltage on the anode (positive terminal). The resulting gas ions are then accelerated and ejected from the thruster by a combination of negatively and positively charged grids with small aperture holes. Typically, ~1,000 to 2,000 volts is presented between these grids. The beam of ions emerging from the thruster has an exhaust velocity between 20 and 50 km/s (Figure 3.15).

If the exhaust was left as a beam of ions, the charges would tend to be attracted back to the spacecraft, the effect of which would be to cancel out the thrust produced. Hence the beam is discharged by electrons from another cathode sited beyond the grids.

The ions are accelerated by the electrostatic force due to the voltage between the plates. However, as suggested by Newton's third law, the ions in turn exert a force on the grids, which is what gives thrust to the spacecraft.

A large fraction of the energy needed to run an ion drive goes into the ionisation process. The choice of propellant is then ideally made from a combination of low ionisation energy, high mass, lack of erosion of the mechanism during use and lack of contamination risk to the vehicle. A current favourite is xenon for its balance of factors. However, it is in short supply and expensive. Alternatives such as bismuth, iodine and argon show promise within the specific requirements of various designs.

The Solar Technology Application Readiness (NSTAR) ion thruster ran in testing for over 30,000 hours of continuous use, mostly at full power, without showing serious signs of wear. The engine works at a power level between 0.5 and 2.3 kW and produces 19 mN of thrust at a specific impulse of 1,900 s in low power mode, rising to 92 mN and 3,100 s at high power level.

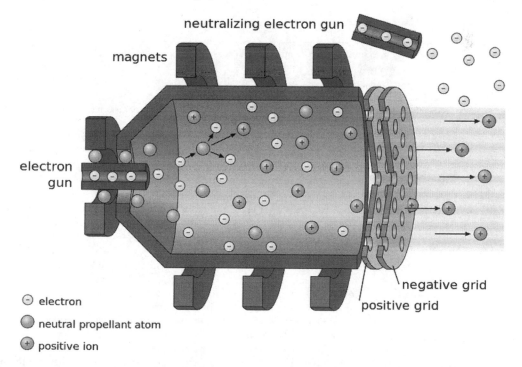

Figure 3.15 A basic schematic of the Gridded Electrostatic Ion Thruster. (Image courtesy of Oona Räisänen via Wikimedia Commons, under Creative Commons Attribution-Share Alike 3.0 Unported Licence. https://creativecommons.org/licenses/by-sa/3.0/deed.en)

In October 1998, NASA launched DS1 (*Deep Space* 1) with the primary mission objective to validate a production NSTAR. Its scientific aim was to return data from a flyby of an asteroid (designated 9969) and a comet (19P/Borrelly). Initially, the drive cut out after 4.5 minutes of operation. The failure was attributed to the grids being shorted by debris from stage separation. The material eventually cleared, probably due to degradation from electrical arcing and/or simply drifting loose. After that, the engine performed well, yielding a total ΔV for the mission of 4.3 km/s while using less than 74 kg of propellant. Electrical power for the probe came from another new technology, which used lenses made from silicon to focus sunlight onto solar arrays.

The *Dawn* space probe carrying three ion thrusters developed from NSTAR was launched in September 2007. Its mission was to visit two asteroids (technically protoplanets) in the asteroid belt. *Dawn* went into orbit around *Vesta* in 2011 and spent 14 months in its company before moving on to *Ceres*, arriving in orbit there in 2015. It currently remains in orbit around *Ceres*. During the mission, the ion thrusters achieved a ΔV of 11 km/s – the record for the largest ΔV achieved by any spacecraft using its own propulsion after separation from its launch vehicle.

NASA is currently developing the Lunar Orbital Platform-Gateway, a proposed space station in the vicinity of the Moon (see Section 9.4). The station will probably utilise the Advanced Electric Propulsion System (AEPS) currently under development by NASA and Aerojet Rocketdyne. The AEPS technology is slightly different from that of the Gridded Electrostatic Ion Thruster as the accelerating cathode consists of circulating electrons trapped by a combination of electric and magnetic fields. Hall Effect Thrusters of this type have been in use since 1971 when the Russians first launched one on a satellite. They have been used routinely for orbital insertion and station keeping ever since. The European Space Agency's 2003 Small Missions for Advanced Research in Technology-1 (SMART-1) spacecraft orbited the Moon for just under 3 years as a technology demonstrator and used a Hall effect thruster.

The propulsion module of NASA's Gateway will be capable of generating 50 kW of electricity. If it has four AEPS engines, they would consume most of the electrical power as each engine can run up to 13.3 kW.

Clearly, ion drive is a valid and developing technology capable of significant performance in the right applications.

ENDNOTES

1. When the first V2 crashed into London, von Braun remarked to a colleague that 'the rocket worked perfectly except for landing on the wrong planet'.
2. During the war, England argued that the Germans were unable to build a rocket due to this weight problem. They did not accept at the time that liquid-fuelled rockets were possible.
3. The 'rocket's casing' includes the structure, engines, propellant tanks and other components that constitute the mass of the rocket *without* counting the propellant onboard and the payload.
4. This is known as the *Venturi effect*. Old-fashioned car fuel systems used carburetors that passed fuel through a narrow (Venturi) throat, reducing the side pressure exerted. Air drawn in from the side started the fuel–air mixing process. Similarly, when air flows over the top of an aeroplane wing, the molecules travel parallel to the surface so the pressure above the wing is reduced. Lift is generated by the greater pressure below the wing, pushing the plane upward. The Venturi effect is the reason people on a train platform are warned not to stand too close to the edge when a train is passing. A train drags air along its path so molecules travel parallel to the tracks. Consequently, the sideways pressure near the train is reduced and people can be pushed forward as the train passes due to the greater pressure behind them.
5. This always sounds terribly painful to me.
6. Sound waves are variations in pressure moving through a gas. If molecules move more quickly than pressure variations can move, the gas will no longer be sensitive to such variations.
7. Rocket engineers speak loosely of converting thermal energy into kinetic energy in an engine bell. Of course, the thermal energy of the gas is a form of kinetic energy – the kinetic energy of the random motion of the gas molecules. They are suggesting that the bell converts the random motions of the exhaust molecules into coherent motion along the line of flight. The pressure of the gas arises from random motions of molecules. When the engine bell is 'tuned' correctly, the gas pressure will equal the external atmospheric pressure. People may wonder how the exhaust exits the bell! The point is that the molecules are released in a straight line (coherent motion); their movement is not 'pressure' in the true sense. There is a subtle difference between random motions and coherent motions of molecules. Pressure in the atmosphere is a result of random motion; wind is generated by coherent motion.

8. This section does not cover the effects of air resistance (drag) due to the flow of air over a rocket in flight.

9. Both companies made good progress and development was equally advanced for both designs. NASA's decision may have been based on the ability of STL to commit more resources and manpower to the project. Rocketdyne was already involved in F-1 and J-2 development.

10. In practice, the engine would be used to increase speed for the first half of the journey and turned around to reduce speed for the second half.

11. The problem is finding a sufficiently long electrical lead.

12. This is why disposing of nuclear waste products by depositing them in rockets and firing them towards the Sun is not a good idea.

Intermission

1

The Saturn V booster rocket

At the time of its construction, Saturn V was the most powerful rocket ever assembled by the United States. More powerful boosters were being tested by the Soviets, but they had an alarming tendency to explode. The Saturn V flew 13 times between 1967 and 1973, including test flights, and still holds the record as the heaviest and tallest rocket ever to become operational. It also lifted the largest payload ever into orbit, although this record depends on the precise definition of payload.[1]

Standing 111 m from base to tip, Saturn V was an enormous structure (Figure I1.1). The rocket comprised three independent stages with their own engines and fuel tanks, two interstage sections with small rocket motors for in-flight corrections, an instrument section with control electronics for the whole stack, and finally the Apollo spacecraft. Standing on top of the command module was the emergency escape system. If a problem developed during the launch, the rocket motors in the escape tower would fire and lift the entire command module clear, to descend to the sea on its parachutes. Alan Shepard, the first American in space, must have faced a sobering moment as commander of Apollo 14 when he entered the command module: the escape tower that would be his life line in case of an emergency was more powerful than the rocket that first lifted him into space!

The fairings at the bottom of the first stage served to smooth the flow of air over the engines in flight. They also contained the retro-rockets that fired to slow the spent stage after the second stage took over. The fins were another aerodynamic aid for the rocket.

The Saturn V was developed by a team at the Marshall Space Flight Center under Wernher von Braun.

One reason that Saturn V achieved its remarkable reliability was the testing and design procedure followed. A great deal of work went into determining the environment in which each part would function. For example, the engineers had to evaluate detailed data about the accelerations, stresses, loads, vibrations, temperatures, pressures, humidity levels and potential material fatigue to which the components would be subjected. Each part was then tested under harsher conditions and required to demonstrate a reliability of 99.99998%. That incredible level was needed as the overall reliability of a system is the product of the reliability factors of all its component parts, and the design criteria for Saturn V called for an overall reliability of 99%. To increase reliability, redundant parts were used where possible.

Assembly of the rocket was a tremendous enterprise in its own right. NASA constructed the giant Vehicle Assembly Building (VAB) to accommodate a rocket standing upright on top of a mobile launcher and caterpillar-tracked crawler that would move it to the launch pad. The VAB stands 160 m high and is one of the largest volumes enclosed under a single roof in the world (Figure I1.2).

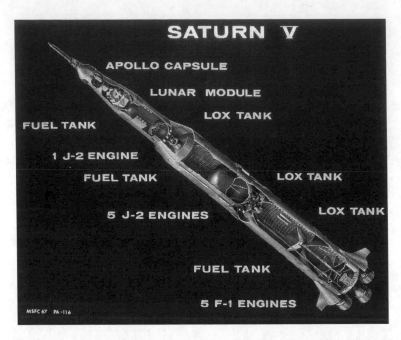

Figure I1.1 The Saturn V Moon rocket and its component stages. (Image courtesy of NASA.)

Figure I1.2 The Apollo 11 Saturn V on its mobile launch platform atop the crawler rolls out of the Vehicle Assembly Building. The smaller building at front centre is the launch control centre. The VAB was utilised during shuttle operations and has been refurbished to accommodate for future operations. (Image courtesy of NASA.)

During the Mercury and Gemini programmes, rockets were assembled, tested and fuelled on their launch pads – a system that had significant disadvantages: the rocket was exposed to the vagaries of weather, and blocked use of the launch pad by other vehicles for long periods. When it was time to design the launch complex for the Apollo Missions, NASA decided to pioneer a new technique based on a mobile launch pad. Two permanent launch pads (39A and 39B) were constructed along with three mobile launchers for transporting rockets to the pads.

Manufacturers delivered the Saturn V components to the VAB where giant cranes capable of lifting 200 tonnes stacked them vertically[2] onto a mobile launch pad (Figure I1.3). The mobile launchers were two-story

Figure I1.3 The Apollo 11 command and service module—lunar module combination is lowered onto the Saturn V third stage inside the vertical assembly building. The lunar module legs can just be seen above the instrumentation unit at the top of the third stage. (Image courtesy of NASA.)

steel structures 7.6 m high, 49 m long and 47 m wide. While resting in the VAB, they were supported on six steel pedestals 6.7 m high. A 121 m umbilical tower stood at one end of the launcher's base. It acted as a frame for the nine hydraulically activated service arms that carried propellants and power lines to the spacecraft and provided access to the rocket for technicians and ultimately, the crew. When a mobile launcher arrived at the assigned permanent launch pad, the umbilical arms were connected to nearby propellant storage tanks and generators. Some of the arms were moved aside in the minutes before launch; the rest automatically swung away as the rocket lifted off. The rocket itself sat over a 14 m square hole in the launcher through which the engine vented into a large trench running under the launch pad.

After a rocket was assembled and tested and the Apollo installed, the next step was transporting the stack to the launch pad. NASA engineers considered various methods for doing this, including the use of a large barge (the first and second stages were transported to Kennedy in this manner). Eventually they decided on driving the mobile launcher and rocket to the launch pad.

Marion Power Shovel Company designed and constructed the crawler. This giant machine carried the full rocket stack and launcher the 5.6 km distance to pad 39 at 1.6 kph. Riding on eight caterpillar tracks, each 2 m wide and 12.5 m long, the crawler was equipped with a sensing system that kept the rocket within 1 degree of vertical (Figure I1.4). A set of 47 m long pipes formed a giant X under the top plate connected to hydraulic jacks at each corner. If the top moved by as little as 1 cm, the jacks could level the system. This was especially important during the climb up the 5 degree slope to the launch site.

The crawler drove into the VAB and under the mobile launcher. The jacks rose to lift the launcher slightly above the pedestals on which it sat, allowing the crawler to drive out of the building with the launcher on top. The reverse process was used to install the launcher on similar pedestals at the pad.

Figure I1.4 The Apollo 11 Saturn V is driven up the incline to the launch pad. Note how the crawler holds the mobile launch pad horizontal despite the incline. (Image courtesy of NASA.)

Each Apollo launch pad was constructed of heavily reinforced concrete and was 0.65 km² in size. The propellants needed for the various stages were stored near the pad, and pumped to the rocket during tests or the final count down. Just before launch, a 590 tonne flame deflector travelled on rails under the mobile launcher and was raised into position by hydraulic rams. Its purpose was to absorb the exhaust blast and deflect the flames along the line of the trench. The deflector was covered with 11 cm of refractory concrete made from a combination of volcanic ash and calcium aluminate (for binding the ash particles together). During a launch, 2 cm of the material was worn away.

A water deluge system cooled the combination of pad, mobile launcher and flame trench. The deck of the mobile launcher was soaked by 29 water nozzles for the first 30 s after launch at a volume of 189,000 litres per minute, after which the rate decreased to 76,000 litres per minute for a further 30 s. The flame trench was cooled at 30,000 litres per minute starting 10 s before liftoff.

I1.1 A TYPICAL SATURN V LAUNCH

Propellant loading of the Apollo spacecraft (i.e. the CM and LM) was performed on the pad prior to launch day. Aerozine 50 served as the fuel and nitrogen tetroxide as the oxidiser.

The process of readying the Saturn V began the day before launch by loading the first stage with its RP-1 fuel pumped through a 15 cm duct at the bottom of the 768,000 litre fuel tank. The fill rate was 760 litres per minute until the tank was 10% full, after which the rate increased to 7,600 litres per minute. Helium was used over the fuel to maintain the pressure in the tank at launch and during flight. Without this, it would have been difficult to ensure a constant flow of fuel to the engines. The helium for this purpose was stored in four 0.88 m³ bottles inside the LOX tank.

The cryogenic propellants (LOX for all three stages and LH$_2$ for the second and third stages) started loading 7 hours prior to launch. If the tanks were at normal temperature, most of the liquid would have boiled into a useless gas as soon as it touched the metal. Consequently, all the cryogenic tanks and pipes were pre-cooled by pumping cold gas through the supply lines. The pre-cooling of one tank could take place while another was being filled.

To avoid damage from LOX splashing about inside, the first stage tank was filled at a comparatively slow rate of 5,600 litres per minute. Even with the pre-cooling of the tanks, much of the LOX loaded initially would boil away. Once the tank was about 6.5% full it cooled sufficiently to prevent any further liquid boiling as it arrived. At this stage, a visual inspection was conducted to ensure the tanks had no leaks. If all was well, filling continued at 38,000 litres per minute until the banks were 95% full, after which the fill rate was reduced back to the earlier slow rate to top up the final load of 1.25 million litres. While the rocket was sitting on the pad, LOX would continually boil inside the tanks and the vapour had to be allowed to escape into the atmosphere. The tanks remained connected to the LOX supply until 160 s before launch so that the contents could be continually topped up.

The second stage tanks were filled at 1,900 litres per minute for LOX and 3,700 litres per minute for LH_2 up to the 5% level, after which filling continued at 19,000 litres per minute and 38,000 litres per minute, respectively. As with the first stage, the tanks were continually topped up – LOX until 60 s prior to launch and LH_2 until 70 s before launch.

In total, it took 4½ hours to load the rocket with its cryogenic propellant. Splitting the sequence allowed checks to be performed and the rocket to settle under the extra stresses produced by the masses of propellant.

The astronauts entered the spacecraft ~160 minutes before launch.

Just 45 s before engine ignition, helium was pumped through the LOX filling lines to pressurise the first stage tank sufficiently to start the engines and build thrust. After liftoff, when the supply of helium was no longer available, gaseous oxygen was used to top up the gas in the LOX tank. Without this, as the volume of liquid in the tank reduced, the pressure of gas over the liquid and hence the flow of LOX would drop.

Liftoff commenced with a start signal to the five F-1 engines of the first stage. The arrangement of the engines across the base of the stage mirrored the pattern of dots on a number 5 domino. To balance and progressively increase the thrust, the centre engine ignited first and then the opposite pairs fired up at 300 ms intervals. The engines were started by opening valves to allow LOX to enter their thrust chambers. A hypergolic solution was then injected into the chambers where it reacted with the LOX. The main fuel valves opened and the RP-1 fuel arrived at the thrust chambers to ignite and sustain the reactions started by the hypergolic solution. At this point, engine thrust would build rapidly. However, the rocket remained attached to the launch pad until each engine developed full thrust. Large clamps then released the stage. As the rocket rose, it drew tapered pins through holes in the launcher with the friction providing some restraining force, so that liftoff was a comparatively gentle affair even though this 'soft release' mechanism restrained the rocket for only ~500 ms.

The Saturn V climbed vertically off the launch pad to an altitude of ~131 m, taking it was well clear of the support arms from the gantry. It then began to pitch and roll into the correct attitude for the rest of the flight. The four outboard engines were mounted to the stage on large universal joints that allowed the whole engine to pivot (gimbal) up to 6 degrees in any direction. The flight path was constantly monitored and corrected by gimballing the engines.

Some 69 s after liftoff, air resistance acting on the rocket reached a peak as the vehicle approached the speed of sound. The drag acting at that moment was up to 2,000,000 N. After the craft passed the sound barrier, the astronauts reported that the ride smoothed out and became much quieter. Most of the LOX and RP-1 was consumed by 135.5 s into the flight. The centre engine was shut down while the remaining four continued to burn until all the propellant was used. Stage separation was achieved by a small explosive charge. Directly after separation, eight small rockets mounted on the first stage fired to decelerate it and allow the second stage to move ahead. The momentum of the first stage carried it up to an altitude of 111 km before it fell back to Earth and crashed into the Atlantic Ocean about 560 km from the launch point.

Meanwhile, eight ullage motors on the second stage fired to settle the propellant so that the pumps could work effectively. Once the engines of the second stage reached 90% of their full thrust, the interstage ring (covering the engine bells and forming a connection to the first stage), separated. This was an extremely precise manoeuvre, as the 5 m tall ring had to slide past the engine bells with a clearance of 0.9 m.

About 30 s after second stage ignition, the escape system was jettisoned from the top of the command module.

The second stage then burned for 395 s, boosting the Apollo nearly into orbit. Once its propellant was exhausted, it separated in a similar manner to the first stage. Four solid fuel rockets mounted around the interstage assembly assured a clean separation of the third stage from the second by slowing down the spent booster. The third stage took over after two solid rockets fired to settle the propellant in its tanks. That stage burned for 165 s, depositing the spacecraft into orbit.

I1.1.1 STAGE DETAILS FROM THE PRESS KIT ISSUED PRIOR TO APOLLO 11 LAUNCH

First stage

Height	42.06 m
Largest diameter	10.06 m
Fully fuelled weight	2,278,247 kg
Empty weight	130,975 kg

Five Rocketdyne F-1 engines burning RP-1/LOX propellant producing a total thrust of 33,375,000 N, uprated over time to a final thrust of 35,155,000 N. The stage burnt for 170 s and carried the rocket to an altitude of ~61 km at a speed of 9,654 kph.

Interstage unit mass	4,583 kg

Second stage

Height	24.84 m
Largest diameter	10.06 m
Fully fuelled weight	480,432 kg
Empty weight	36,250 kg

Five Rocketdyne J-2 engines burning LH_2/LOX propellant produced a total thrust of 4,450,000 N. The second stage burnt for 395 s to an altitude ~184 km and speed of 24,617 kph.

Interstage unit mass	3,665 kg

Third stage

Height	17.77 m
Largest diameter	6.61 m
Fully fuelled weight	118,171 kg
Empty weight	11,340 kg

A single Rocketdyne J-2 engine burning LH_2/LOX propellant produced 890,000 N of thrust. The third stage's initial 165 s burn to place Apollo in Earth orbit at an altitude of 185 km and 24,617 kph. The third stage was lit again to inject Apollo onto an orbit that would carry it to the Moon. This Translunar Injection burn lasted 312 s and lifted the spacecraft's speed to 39,420 kph.

Instrument unit mass	1,953 kg
Total Saturn V mass fuelled	2,887,051 kg

I1.1.2 PAYLOAD CAPACITY (TOTAL MASS OF COMPONENTS ABOVE INSTRUMENT UNIT)

Apollo 8's Saturn V	36,000 kg
Apollo 17's Saturn V	53,000 kg

These two figures demonstrate the soundness of the Saturn V design and how the engineers were able to increase the available thrust over time.

I1.1.3 SATURN V CONTRACTORS

First stage	Boeing Company
Second stage	North American Aviation (Rockwell International)
Third stage	Douglas Aircraft (McDonnell Douglas)
F-1 engines	Rocketdyne
J-2 engines	Rocketdyne
Instrument unit	IBM and Marshall Space Flight Center

ENDNOTES

1. If you define the payload into orbit as the Saturn V third stage, the CSM–LM stack and unused fuel in the stage, the record goes to Apollo 17 with an orbital payload of some 141 tonnes. The first American space station, *Skylab*, was converted from a Saturn V third stage and weighed 88 tonnes.
2. It is said that the crane operators were capable of lowering an 40 tonne weight onto an egg without cracking its shell.

4

Orbits and trajectories

4.1 HOLLYWOOD GETS IT RIGHT

Over the years, writers of science fiction films have used many literary devices to overcome the basic problem of trying to film people who are supposed to be in space: how to simulate the lack of gravity. Some scripts barely mention the problem in passing through vague references to artificial gravity – a nice idea, but with no conceivable basis according to our understanding of physics. Others have made reference to revolving sections of the spacecraft used to simulate gravity inside.[1]

However, a director trying to film the true story of what happened onboard a real spacecraft would have to deal with the problem head-on. This was the dilemma facing the producers of the film *Apollo 13*. Most events in the story take place within the Apollo command and lunar modules, which did not spin fast enough to produce any noticeable simulation of gravity inside.[2]

The solution in *Apollo 13* was to make use of a standard NASA training technique: fly the crew in an aeroplane and use a skilled pilot to put the plane into *free fall*. The aeroplane is flown on a specifically curved path known as a *parabola*. At the peak of the curve, the plane dives towards the ground, still following the shape of the curve, which is exactly what would happen if the plane fell freely from a great height (Figure 4.1).

As a result, any item that is not tied down inside the plane starts to float about. For the movie, various spacecraft sets were constructed inside the plane and filming took place during each run. In this manner, the director got a realistic simulation of the astronaut's experiences on the way to the Moon. However, while the aeroplane followed the parabolic curve, its crew, the actors and all the equipment onboard were still under the influence of the Earth's gravity. This seeming paradox is resolved by fully understanding what is meant by 'free fall' and 'weightlessness' and how objects can be placed in orbits that will circle a planet indefinitely.

4.2 FALLING AGAIN

Free fall is the physicist's term for falling with only the force of gravity acting. If you were to jump off a diving board into a swimming pool, you would experience a pretty close approximation to free fall on the way down. The air resistance on your body would be small, as you would not be moving very quickly. However, were you to jump from an aeroplane, as part of a skydiving team perhaps, then after a few moments, you would no longer be in free fall. From skydiving altitude, the speed at which you fall would build up considerably (you have further distance and more time to fall) and so would the air resistance. Skydivers spend most of their time before opening their parachutes falling at a constant speed, as the air resistance is equal to the pull of gravity on them: this is known as falling at *terminal velocity*.[3]

Section 2.5 explains how specially designed couches provide the force required to accelerate astronauts during liftoff. The idea was illustrated by comparing a spacecraft couch to a weighing machine composed of a large spring on which an astronaut stands. The device measures weight by recording the amount of compression in the spring. As a rocket takes off, the spring is caught between the force due to the astronaut's weight acting down and the force of the engines transmitted upward through the floor. Consequently, the machine records greater weight than if the rocket was stationary on the ground. We experience something very similar when we stand in a lift that accelerates upward.

Now consider what would happen to the machine if the rocket were falling towards the ground. For the moment, forget about air resistance and assume that the rocket and its contents are in free fall.

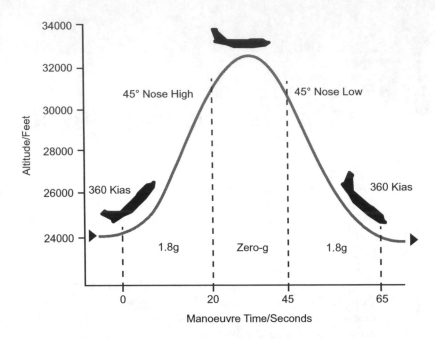

Figure 4.1 The trajectory followed by a NASA training aeroplane showing the g forces experienced at each stage of the flight. (Image courtesy of NASA.)

In that case, we can say that the astronaut, the spring and the floor are *falling towards the ground at exactly the same rate.* In Section 2.4, we discussed Dave Scott's demonstration on the Moon and the surprising fact that, in the absence of air resistance, all objects have the same acceleration under gravity. As a result, our falling astronaut is not really in contact with the top of the spring. He is falling towards the ground just above the spring.

In a stationary situation, the spring is compressed as the astronaut is pulled towards the ground by the force of gravity. In turn, the spring is held up by the rigid floor, so it is forced into compression as it is trapped between two opposing forces. In free fall, both the astronaut and the spring are falling and they keep exact pace with each other, even though they have very different weights. The spring is not being compressed, so the measuring component of the device would read zero. The astronaut certainly still has weight, in the technical sense of being pulled towards the Earth by the force of gravity. However, all the physical sensations that go along with our experience of weight will have disappeared.

Physicists define weight as the force of gravity acting on an object. This force always acts at the centre of gravity, which, for humans, is in the middle of the body around the height of the navel. As people do not have a sensory organ that detects this force, when we talk about weight in common terms, we must be referring to something other than the very specific force that physicists have in mind.

At the moment I am sitting in a very comfortable chair that my family bought for me some years ago. My immediate experience of weight is the sensation of the chair pushing up on me. I also have some feeling that my lower legs are hanging down and pulling on my knees (the chair is tipped back slightly so my feet do not quite touch the floor). The tendons and muscles near my knees can detect forces acting on them. They must extend or contract to balance the forces. The various sensations related to this physical activity are what we associate with having weight.

A person standing on the ground is in a state of compression, just like the spring. They are pulled towards the ground by gravity and pushed upward by the ground beneath their feet. The person's joints and bones are being compressed. In free fall, only the force of gravity is acting, so the body is not in compression. The joints open out slightly and the sensation of weight disappears. This is what we mean when we say that an astronaut in orbit or in free fall in a training aeroplane is weightless. Weightlessness, in this context, does not mean an absence of gravity. Indeed, no human has ever been in a situation where no force of gravity was acting.[4]

Inside a spacecraft in orbit, an astronaut could float weightless, barely in contact with the floor. If she held an object in front of her and let go, the object would float, precisely because it is falling towards the ground at exactly the same rate as the astronaut and the spacecraft.

While travelling towards the Moon, the astronauts are still in free fall. With no engines in operation, the only force acting on them is gravity and so they are still falling. For a while they are falling back to Earth (although still getting further away), then as they get nearer to the Moon, they start to fall in its direction. It seems odd to talk about falling towards the Earth while travelling away from it. However, think about throwing a ball up in the air. While it is moving up, the Earth is pulling it back.

Now that we have developed this understanding, we can see how it follows that a spacecraft cannot remain in orbit *unless the force of gravity is acting*. If, as you might suspect from the expression 'weightless', there was no force of gravity involved, then, according to Newton's first law, the spacecraft should move in a straight line with a constant speed. Such a path would take it away from the Earth and into space.

A spacecraft in orbit (and all the people and objects within it) is in a continual state of free fall. Above the atmosphere, there is no air resistance, no matter what the speed. The entire craft and its contents are constantly falling towards the ground but never actually get there.[5]

4.3 ORBITS

At the end of the last Apollo 14 moonwalk, mission commander Alan Shepard connected his sample collector to the head of a 6-iron golf club and, in front of a bemused TV audience, hit the first golf shot on the Moon (one-handed as his bulky spacesuit did not allow him to get a proper grip on the club) (Figure 4.2).

Figure 4.2 The view from the Apollo 14 lunar module at the end of the final extravehicular activity. Alan Shepard hit a golf ball by attaching a 6 iron to his sample collector. The ball did not go very far (probably because it is difficult to swing a club in a bulky spacesuit) and can be seen in the ring (added by the author) on the image. The pole next to the ball is a 'javelin' thrown by a crew member. (Image courtesy of NASA and the Apollo 14 crew.)

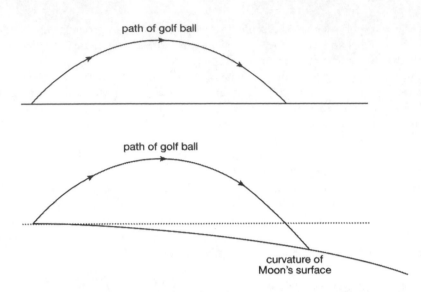

Figure 4.3 To calculate the true distance travelled by a golf ball on the Moon, the curvature of the surface must be taken into account.

The path followed by a golf ball in the absence of air resistance, and without spin to complicate the issue, is a *parabola*. Any object that moves horizontally at a constant speed and accelerates vertically at a constant rate will follow a parabolic path.

Such a curve is shown in Figure 4.3. The top picture shows a golf ball in flight on the Moon. It is a simple matter for those familiar with the physics involved to calculate how far the ball would travel, given the speed at which it left the club and the angle at which it started to fly. However, this would not yield a true distance unless the curvature of the ground was also taken into account.

Earth is a large planet, so the curve of its surface is quite small over a region the size of a golf course. However, someone firing projectiles from one part of the world to another, or at least planning to fire them, must take the curvature into account. As is clearly shown in the lower illustration in Figure 4.3, the ball will actually cover a slightly greater distance as the ground is not flat. In fact, strictly speaking, the path of the ball is not a parabola. This is because the pull of gravity is always towards the centre of the Earth, and so the direction of the pull changes during the flight. For the path to be an exact parabola, the direction of the force must not change. However, for our discussion, we will ignore this complication and still refer to parabolic paths.

Figure 4.4 shows three paths. The ball is hit progressively harder, but at the same angle to the ground in each case. The faster the ball leaves the face of the club, the less curved the path and the further it will travel.

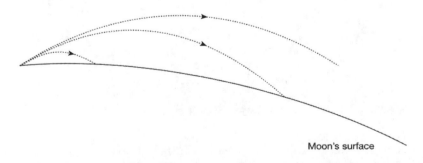

Figure 4.4 Striking a ball at the same angle, with different speeds.

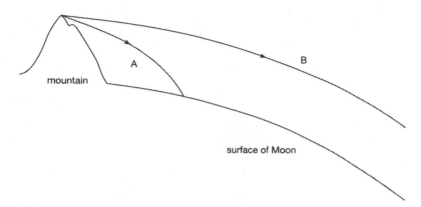

Figure 4.5 A ball putted off the side of a mountain will fall to the surface if hit too slowly (A) but will be sent into orbit if it reaches the correct speed (B).

The next step is to change from a 6 iron to a putter and imagine a putt from the top of a mountain (Figure 4.5). If the ball is struck horizontally (parallel to the ground), it will roll off the top of the mountain and fall in a path that corresponds to part of the equivalent path for a 6-iron shot. If the planet were flat, it would be half of a parabola.

As with the 6-iron experiment, the faster the ball moves, the less steep the curvature of the path. In fact, when a ball moves sufficiently quickly, *the curve of its path matches the curve of the planet's surface*. When this happens, the ball will fall continually towards the surface, but the surface will, in turn, curve away from the ball along a parallel arc. The ball falls without getting any closer to the surface. The ball is in orbit and in free fall. If the same situation were reproduced in a spacecraft, it would also be in free fall and all its occupants would be 'weightless'.

In practice, putting a golf ball into orbit from the side of a mountain would be impractical, even on the Moon. At such a low altitude, it would be very difficult to avoid hitting other mountains. The point of this golf example is to illustrate what an orbit is, not to suggest a practical procedure. Newton used a very similar argument – using a cannon to fire projectiles from the side of a mountain. In reality, rockets launched from the surface follow a specific trajectory into orbit. By the time they reach the chosen altitude, they are moving parallel to the ground and at the correct speed. At that point, gravity will continually pull the spacecraft towards the planet in a constant orbit.

4.4 CENTRIPETAL FORCES

Much of our common experience indicates that a force pushes outwards on a person or object turning on a curved path. For example, if you are sitting in a car that takes a corner sharply, your body tends to sway outward, away from the centre of the circle. Despite the powerful persuasion of such experiences, no *centrifugal* (outward pointing) force acts on you as the car turns. In fact, you move *outward* as there is insufficient force pulling you *inward*.

When a car turns a corner, a force must be applied to change the direction of motion (Newton's first and second laws). A moment's thought about the way in which the direction of travel is changing should convince you that the force must be pulling *inward* towards the centre of the circle, as that is the direction in which the path is changing. This force is provided by the friction between the tyres and the ground. A similar force must act on a person sitting inside the car if she is to follow the path of the car. The force must be provided by the friction between the person's clothing and the car seat. If the force is insufficient, her body will follow a straighter path than the car. This will move her outward relative to the car, as the car turns beneath her body. Eventually, the side of the car or the side of the seat will come to her aid and provide an additional force to push her inward with the car.

In orbit, the only force acting on any object placed there is gravity and as a result, all objects follow identical paths (unless they were knocked about inside a spacecraft). Gravity provides the *centripetal* (inward pointing) force required to maintain a circular orbit. In general, the size of the force needed to maintain a circular path depends on the mass of the object, the speed at which it travels and the radius of the circle.

- The more massive the object, the greater the force required. This is a direct consequence of Newton's second law of motion.
- The faster an object is moving, the greater the force required. This is also a consequence of the second law.

Remember that Newton's law is best expressed by saying that the force required is equal to the rate at which the momentum changes. Momentum equals mass times velocity, so both factors influence the size of a centripetal force.

When considering circular motion, it is often easier to work in terms of *angular speed* rather than ordinary *linear speed*. Angular speed is the angle through which an object has turned divided by the time it takes to perform this manoeuvre. In everyday life, angular speeds (rates of rotation) are most often specified by the number of revolutions per minute. For example, we quote the revolutions of an engine in RPM (revolutions per minute). Vinyl long-playing records spin on their turntables at 33 RPM. CDs can spin at rates up to 500 RPM.

Physicists are not very comfortable with terms such as 'revolutions' (which do not make their equations look very pretty) or minutes (not a standard unit). They prefer to use *radians per second*. Those unfamiliar with the radian unit of angle can review the explanation below (see the box 'Radians').

Radians

Most of us are familiar with the degree as a measure of an angle. After all, this is the unit used on the protractors that we first came across early in our school careers. Protractors work by dividing a full circle in half (a cut across the diameter). The remaining half circle is then divided up by marking 180 equally spaced dashes along the circumference of the semicircle. When these dashes are connected to the centre by drawing lines along the radius, the angle between two successive lines is defined to be 1 degree.

Dividing a semicircle into 180 segments (360 for a full circle) is an arbitrary choice. It seems entirely natural, simply because most of us have used the 360 degree measure since we were teenagers. Yet it is a choice, and not an especially sensible one, as it is unrelated to any property of circles. This would not matter much if there was no sensible alternative (if you must make a choice between equally arbitrary alternatives, then 360 is as bad as any other), but there is one. Every circle has a circumference equal to 2π times its radius. This suggests that it would be far more natural to divide the circumference into 2π equal pieces. The snag is that π is not a whole number (3.1415926535897932384 …). In practical terms, a protractor could not be made to show this radian measure of an angle. However practical matters tend not to stop mathematicians. It is a simple matter to convert from degrees to radians and back again. One complete circle (360 degrees) equals 2π radians. From this it follows that 180 degrees is equal to π radians, and 90 degrees is $\pi/2$ radians. By simple proportion:

$$\theta \text{ degrees} = \frac{\theta}{360} \times 2\pi \text{ radians}$$

In physics, angular speeds are quoted in radians per second (RPS or rad/s, not to be confused with RPM) and given the symbol ω. The advantage of using angular speeds is that they are independent of the distance from the centre of a circle, whereas linear speeds are not.

The Earth rotates on its axis once every 24 hours. This is an angular speed of 2π radians in 24 hours or $24 \times 60 \times 60 = 86,400$ s, which is 72 microrad/s (put that way, the velocity doesn't sound very fast, does it?). In linear terms, however, as the Earth has a radius of 6,400 km, a point on the equator must cover a circle 40,212 km in circumference in 24 hours. That is a speed of 465 m/s.

A spacecraft orbiting the Earth once every 24 hours at a height of 35,850 km would have to move at 3.07 km/s. The spacecraft and the point on the equator have the same angular speed, but their linear speeds are rather different.

The centripetal force required to keep an object pulled into a circular path can be calculated from the equation:

$$F = \text{Mass} \times \text{Angular speed} \times \text{Radius} = m\omega r$$

If we keep the angular rate and the radius constant and try to launch different masses into that orbit, they will require different forces to act on them. In fact, the size of the force will be proportional to the mass of the object.

On the other hand, if we try to launch objects of identical mass into orbits with the same angular rate but different radii, then the force required is proportional to the radius. However, there is one important point to bear in mind. Unless the force keeping the object in orbit is supplemented by engine thrust (or another mechanism; see Section 4.5), gravity will have to do the job alone. We cannot change the force of gravity; it has a particular value at a particular distance from a planet. In that case, *only one speed can be used by an object at a given distance from the Earth if it is to remain in orbit under gravity.*

4.5 GRAVITY AND ORBITS

Given that the force required to keep an object moving in a circle at a constant angular rate *increases* as the radius of the circle increases and that the force of gravity *decreases* the further you get from a planet, it makes sense that gravity can provide the force for a given rate at only one radius. The graph in Figure 4.6 shows how the centripetal force required to keep an object circling the Earth once every 24 hours increases with distance from the centre of the planet. Also shown is the manner in which the pull of gravity decreases with distance. There is one distance at which the two lines cross: where the force of gravity provided by the planet equals the centripetal force required (the figures in the graph were calculated for a 1 kg object).

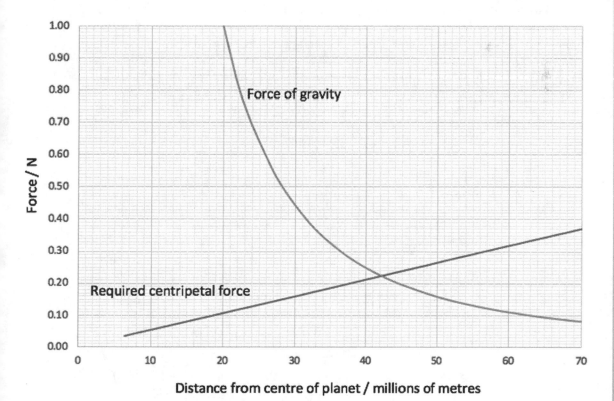

Figure 4.6 Comparing the pull of gravity from Earth with the centripetal force required to keep an object in orbit with an angular speed sufficient to complete the orbit in 24 hours. The lines cross at the distance where a satellite would be able to maintain the orbit without powered assistance.

If you try to place an object in orbit moving at this angular rate but nearer than the crossover distance on the graph, gravity will be greater than the centripetal force required. Consequently, the object will fall towards the Earth. On the other hand, if the object is too far out in space, gravity will be insufficient to hold it and it will tend to fly away. Placed at that single distance, moving at that speed, the object will continue to circle the Earth indefinitely.

The orbit discussed here is quite ordinary in terms of physics, but is unique in its usefulness to people on Earth. As far as anyone standing on the surface is concerned, a satellite placed in this orbit will appear to *hover in a stationary manner above the ground*. In fact, the satellite races around the Earth once every 24 hours. As 24 hours is the time it takes the Earth to rotate on its axis, the satellite is simply keeping up with the ground.

This is known as a *geostationary orbit* and it is the orbit of choice for communications satellites. From our perspective, the satellite remains floating in the sky at the same place relative to the ground so we have no need to 'track' a communications aerial across the sky as the satellite passes. This is why satellite television dishes can be pointed in one direction and left in that position. In essence, they are 'talking' to a satellite in a geostationary orbit over the equator.

Figure 4.7 shows three possible paths around a planet. A satellite placed in an *equatorial orbit* circles around the planet's equator and will be visible directly overhead by people living at this latitude. A *geostationary orbit* is an equatorial orbit in which a satellite turns about the Earth in the same time period that the Earth requires to turn about its axis.

A satellite in *polar orbit* moves from one pole to another. Viewed from Earth, a satellite in a polar orbit does not appear stationary. The radio dishes communicating with it must track its path across the sky. While the satellite moves from north to south (or the other way) the world beneath it turns from west to east. As a result, the satellite will eventually fly over every part of the Earth's surface. This makes polar orbits very handy for spy satellites and for more conventional research satellites that monitor weather patterns, plant growth, land use, and so on.

The unstable orbit shown in Figure 4.7 cannot last, as the force of gravity acting on the satellite pulls it towards the centre of the planet, which is not the same as the centre of the orbit's circle. Such an orbit cannot be sustained without another force such as engine thrust being in action.

As noted earlier, a geostationary orbit is very useful for communications satellites to the point where the orbit is becoming rather crowded. Although a great deal of space remains between the satellites, they require a minimum separation to prevent their signals overlapping with each other on the surface of the Earth. Of course, an infinite number of equatorial orbits exist, but we have only one *unassisted* geostationary orbit.

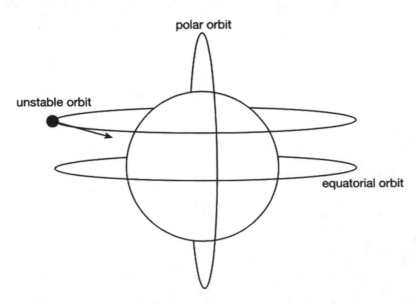

Figure 4.7 Three possible orbits.

In principle, you can set a satellite turning about the Earth once every 24 hours at any distance from the surface, provided you supplement the gravity by other means. Using thrust from engines requires a store of propellant and would mean that the satellite has only a finite useful time in orbit. It is far better to employ a naturally occurring resource: sunlight.

A suitably equipped satellite can use the pressure exerted by sunlight reflecting off a large mirrored surface to provide an upward thrust that helps balance the pull of gravity. In this way, a satellite with a mirror can be placed in a 'powered' geostationary orbit nearer to the Earth than the gravitationally supported orbit. This scenario would require enormous mirrors, but a thin aluminium foil stretched over a light framework forms a perfectly good mirror in space as it does not have to survive wind, rain or other inconveniences that we experience in the atmosphere.

Russia has already launched smaller orbiting mirrors. The *Znamya* satellite in 1993 was used to conduct experiments using reflected sunlight to illuminate areas subject to long twilight, such as the steppes. China is looking to carry out similar plans in 2020 and 2022. The technology for these mirrors would essentially be the same as that needed for a solar-supported geostationary orbit.

4.6 OTHER ORBITS

It is possible to put a mass into orbit at any distance from the centre of the Earth, provided the correct speed is used. Given that the orbit is not power assisted, and that the force of gravity is a fixed quantity at any radius, the speed has to be chosen so that gravity can provide the necessary centripetal force.

As Figure 4.8 illustrates, the farther one is from the centre of the planet (in this case Earth) the slower the orbit has to be. This is because gravity decreases with distance from the gravitating mass.[6] It will be important to remember this relationship later in this chapter when we come to discuss moving from one orbit to another.

In practice, most orbits are neither equatorial nor polar. They are a part mixture. The plane of the orbit still passes through the centre of the Earth but can be oriented at any angle.

4.7 SIMULATING GRAVITY

With a sufficiently large space station or spacecraft, it is possible to use rotation to simulate the effects of gravity. As an example of how this might be done, consider an astronaut with magnetic boots standing on

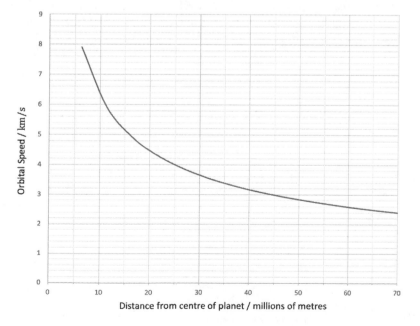

Figure 4.8 How the linear speed needed for a stable orbit decreases with distance from the centre of the Earth.

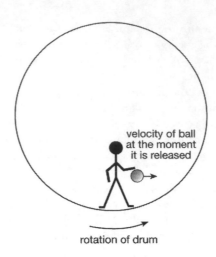

velocity of ball
at the moment
it is released

rotation of drum

Figure 4.9 A gravity simulation experiment.

the inside of a rotating drum which is in space far away from any star or planet so that no gravitational force can act on the drum (Figure 4.9).

The astronaut holds a ball in his hands. While he is holding the ball, his hands exert the necessary centripetal force for the ball to match the rotation of the astronaut (the centripetal force exerted on him comes from his magnetic boots[7]). At the moment he releases the ball, it no longer has any centripetal acceleration. Consequently, it carries on in a straight line with the velocity it had at the moment it was released (assuming that air friction is too small to be of any consequence).

Figure 4.10 illustrates what would happen to the ball. Each part of the diagram represents a small period of time in chronological sequence. The parts have been separated in a horizontal direction to make what happens clearer.

The ball continues in a straight (dotted) line, eventually reaching the side of the drum. The natural tendency is to think that the ball will roll back but remember that the drum is not on Earth. No gravity acts downward in this diagram! From the astronaut's view, the ball has fallen as if pulled by some gravitational force towards the side of the drum. When the ball hits the drum wall, the force of the drum on the ball will provide centripetal acceleration so the ball will keep pace with the rotation from then on. This example is a simulation of gravity, *not* real gravity. A stationary ball placed precisely at the centre of the drum would stay there. There is no force pulling it to the side.

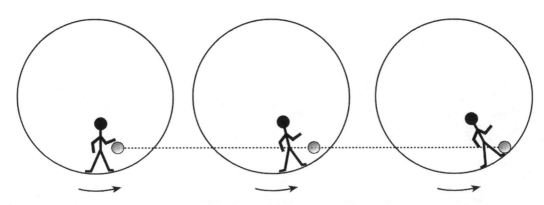

As the drum turns the ball continues to
move in a straight line at constant speed

Figure 4.10 More experiments with simulated gravity.

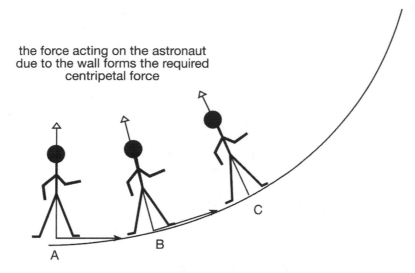

the force acting on the astronaut
due to the wall forms the required
centripetal force

Figure 4.11 An astronaut not wearing magnetic boots would feel a gravity-like effect.

An interesting example of this appeared in an episode of the science fiction series *Babylon 5*. The show was set on a giant space station shaped like a cylinder rotating about an axis along its length. One character was forced to jump from a train car running on a track mounted along the central axis to avoid a bomb explosion. Although the space station was spinning fast enough to simulate full Earth gravity for those living on the interior surface of the cylinder, the character did not fall towards the surface. Instead, the force exerted on him by the train when he jumped clear had given him a forward velocity and he drifted slowly toward the 'ground'. However, this was not a safe situation as the 'ground' beneath him rotated at great speed. The situation was similar to parachuting onto a moving train. The vertical speed would be gentle but the horizontal speed…

Now consider an astronaut without magnetic boots. At the point of first contact with the walls of the drum, friction would establish a forward motion parallel to the direction of the drum at that time. The astronaut would then drift along at constant speed in a straight line until striking the rotating wall of the drum (from A to B in Figure 4.11). At this point, the impact would provide a force. Some of the force is friction and would establish a new velocity for her. The other part of the force results from the inward push of the drum on the astronaut (which is why she does not break through the surface of the metal). This force gives her centripetal acceleration.

In practice, an astronaut would not drift across in the manner described above because the curvature of the drum walls would be far smaller. However, the example illustrates that frictional force couples the astronaut to the drum's rotation, and as the drum rotates, it produces an inward facing force that would compress her joints. Consequently, she feels as if gravity is acting. If the rotation rate was sufficiently quick to produce a centripetal acceleration of 1*g*, the astronaut will feel as if she were standing on Earth. In practice, this would need a large diameter drum, otherwise the rotation rate would be ridiculously high.

4.8 CHANGING ORBITS

One of the most decisive steps in lunar mission planning was the selection of a preferred method from three viable options: (1) launching a rocket from the surface of the Earth straight to the Moon (*direct ascent*); (2) assembling a craft from parts linked together in Earth orbit (*Earth orbit rendezvous*); and (3) requiring a command module (mother ship) to dock with a lunar module in orbit around the Moon (*lunar orbit rendezvous*). As we will discuss in Chapter 5, the final decision involved balancing risks against technical limitations and minimising the mass of the craft. On that basis, lunar orbit rendezvous was selected.

After the decision was made, it became clear that NASA and its astronauts had to become experts in the mechanics of moving among orbits, so that spacecraft could find each other in space and ultimately dock safely. The mission plan also called for the spacecraft to spend some time in Earth's orbit, then follow a trajectory to the Moon and enter a specific lunar orbit that would pass over the landing site. All these manoeuvres would require ΔV changes brought about by firing main engines or thrusters.

To understand how to overcome these challenges, we must discuss the physics of orbits in more detail. Earlier in this chapter we described circular orbits. However, precisely *circular* orbits are rare; most orbits are *elliptical*. An ellipse is described mathematically by using two points in space called *foci* (plural of *focus*). An ellipse is a curve around the foci constrained by the following rule: for any point on the curve, the length of the line from the point to one focus added to the length of a line from the point to the other focus is always the same (Figure 4.12). A circle is an ellipse with the two foci at the same place – the centre of the circle.

Johannes Kepler was the first person to demonstrate that the planets orbit the Sun along elliptical paths. He discovered this by painstakingly tracking the path of Mars (using the observations of Tycho Brahe) and trying to match the orbit to a mathematical shape. Some years later, Newton used his newly discovered law of gravity to prove that all objects must orbit in elliptical paths. For every elliptical orbit, one focus must be the centre of gravity of the attracting object.[8] The other focus is simply a spot in space that does not have to contain anything.

As a satellite loops about the Earth (Figure 4.13), it passes through a point where it is nearest to the ground (the *perigee* of the orbit) and, at the opposite end of a line through the foci, a point where it is furthest from the ground (the *apogee*). For orbits around the Sun, these points become the *perihelion* and *aphelion*, respectively.

The greater the distance between the foci, the more *eccentric* the orbit and the greater the difference between apogee and perigee. A circular orbit has no distinguishable apogee or perigee; it also has zero eccentricity. In practice, many planetary orbits are so nearly circular that a casual glance would be unable to distinguish them from exact circles.

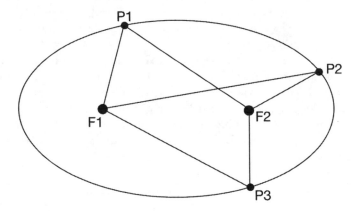

Figure 4.12 An elliptical path. The two foci are marked *F1* and *F2*. At each of the three points (*P1* through *P3*), adding the lengths of the lines connecting the points to the foci always yields the same answer.

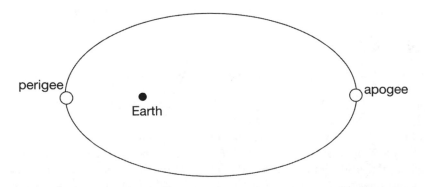

Figure 4.13 An elliptical orbit has a point of lowest altitude (perigee) and one of highest altitude (apogee).

Another factor to be considered when plotting the path of a satellite or spacecraft is the velocity at which the object moves. On a circular orbit, the velocity is constant and always directed along a tangent[9] to the path. However, with an elliptical path, the size of the velocity varies from a maximum at perigee to a minimum at apogee. Also, around an ellipse the direction of travel at any point is not simply tangential. The easiest way of specifying the direction is to split it into two parts: along the line connecting the spacecraft to the centre of gravity of the planet (*radial motion*), and along a line at 90 degrees to the radial motion (*horizontal motion*).[10]

As can be seen in Figure 4.14, the only two points at which the velocity is purely horizontal are at apogee and perigee. A spacecraft passing through its perigee is moving faster than an equivalent craft on a circular orbit of the same radius. Consequently, the force of gravity at perigee is not enough to hold it in a circular orbit. As the craft moves away from perigee, it gains a radial component of velocity pointing away from the centre of the Earth. However, it is not moving fast enough to totally escape the pull of Earth's gravitational field. It loops away, slowing down as it goes (just as a rock thrown into the air will slow down to rest before falling back). At apogee, its radial velocity reaches zero, so the craft again has a purely horizontal velocity. However, now it is moving too slowly for a circular orbit at this radius and the gravitational force pulls it in. Of course, it does not fall directly towards the Earth because it has not lost its horizontal velocity. It loops back towards perigee. The pattern repeats itself unless some other force such as air resistance, if perigee grazes the atmosphere, or an engine burn occurs.

If a spacecraft in an elliptical orbit were to light its engine, the resulting change in the orbital path would depend on the ΔV of the burn, the direction of thrust applied and the point in the orbit at which the burn took place. All these factors are complicated and the general effect on an orbit requires careful calculation. However, it is possible to make some simplifying assumptions to clarify the idea:

- the burn takes place either at perigee or apogee;
- the thrust direction is along the line of flight;
- the burn duration is very short compared to the orbital period.

If the first two assumptions are true, the burn will raise or lower the horizontal velocity of the spacecraft at the points at which it has only horizontal velocity. The third assumption is an abbreviated form of the *impulse approximation*: the ΔV produced by the burn happens instantaneously at a single well-defined point

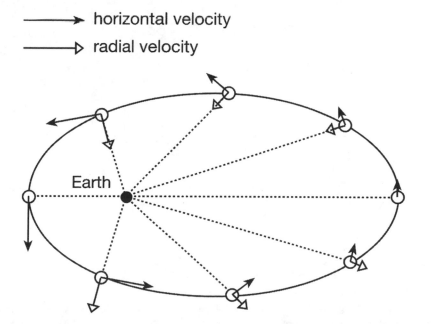

Figure 4.14 The horizontal and radial components of a spacecraft's velocity in an elliptical orbit.

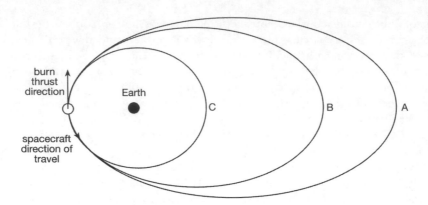

Figure 4.15 Firing the engines at perigee with a thrust direction against the horizontal velocity will slow the spacecraft down. (Remember that the direction of thrust is opposite to the direction in which the exhaust is emitted.) As a result, path A will be turned into a less eccentric orbit (B) or even into a circular orbit (C) if the ΔV is correct.

on the orbital path. Of course, in practice, the ΔV is spread over the burn duration and the spacecraft moves along the orbit during the burn which affects the final orbital change.

Consider a spacecraft in an elliptical orbit which burns its engines at the moment it reaches perigee. The subsequent path it takes will depend on whether the ΔV increases or decreases the velocity. Decreasing the velocity (Figure 4.15) means that the centripetal force required to keep the craft in a precisely circular orbit is lower and closer to the actual force that Earth's gravity can provide. The resulting path is more nearly circular. In fact, with a correct burn, a spacecraft can lose sufficient velocity to change from an elliptical to a circular path with a radius equalling the height at perigee on the old ellipse. This is the process of *circularising* an orbit. Increasing the velocity will have the opposite result. The spacecraft enters a more eccentric orbit but the height and position of perigee remain the same. Apogee moves further away.

Applying a horizontal burn at apogee (Figure 4.16) has similar effects. If the resulting ΔV increases the velocity, the orbit can be circularised with a radius equal to that of the apogee. On the other hand, slowing the spacecraft down at apogee will tend to make the orbit more elliptical.

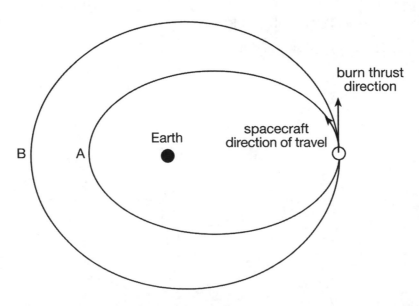

Figure 4.16 A burn performed at the apogee of an elliptical path (such as A) that increases spacecraft speed will tend to raise the height of perigee. With sufficient ΔV, the height of the perigee can be lifted to equal that of the apogee: the orbit becomes circular.

The other simple type of burn is radial. The spacecraft is oriented so that the engines thrust along the line connecting the craft with the centre of the planet (or the focus of the orbit). At any point on the ellipse between apogee and perigee, the craft will have both horizontal and radial velocities. If a burn is timed to yield a ΔV equal to the radial velocity at the moment of burn and pointing in the opposite direction, the radial velocity can be cancelled out. This will convert the orbit into a circular one that has no radial velocity. Similarly, radial burns can be used to turn circular orbits into elliptical ones. Unlike a horizontal burn that makes the position of a spacecraft into the apogee or perigee of a new orbit, a radial burn will move the craft onto an orbit somewhere between apogee and perigee, depending on the ΔV applied.

4.8.1 HOHMANN TRANSFERS

In 1925, Dr Walter Hohmann, a German engineer, demonstrated mathematically that the best way to transfer between two circular orbits was by using an intermediate path that was an ellipse (a *Hohmann transfer orbit*), as this used the least propellant.

Assume that a spacecraft starts in a low circular orbit and the occupants wish to move it into a higher circular orbit (to rendezvous with another spacecraft or satellite, perhaps). They start by performing a burn that accelerates their horizontal velocity. This moves the craft from a circular orbit to an elliptical one. The perigee of the new orbit is the point at which the burn took place (remember we are still using impulse approximation). The ΔV achieved by the burn determines the height of the apogee the craft will reach on the opposite side of the orbit to the burn point (180 degrees around). The astronauts will select the ΔV that places the apogee of their new orbit at the same height as the required circular orbit When they arrive at apogee, they must burn again to circularise their path by accelerating the spacecraft. With the correct ΔV from this second burn, they will settle into their new circular orbit (Figure 4.17).

At first sight, moving from a low orbit to a high one by making two *accelerating* burns appears odd. After all, a high orbit will require less speed for stability than a low orbit. However, as the spacecraft coasts

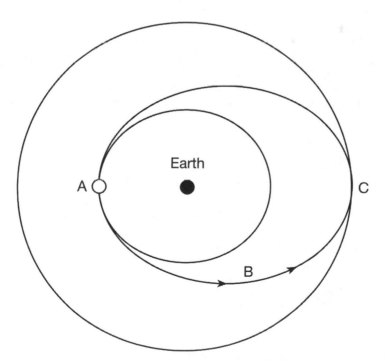

Figure 4.17 A Hohmann transfer from a low orbit to a higher one. A spacecraft at point A ignites a burn that increases its horizontal velocity. The burn point then becomes the perigee of a Hohmann transfer orbit. The spacecraft coasts along the orbit (B) until reaching the apogee (C) which intersects with the higher orbit. Another burn at this point can accelerate the craft, placing it on the higher circular orbit.

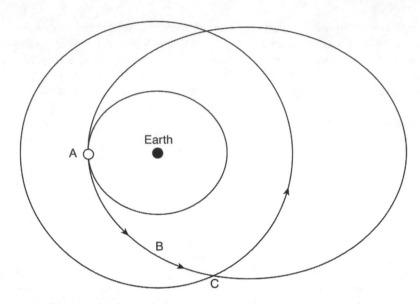

Figure 4.18 This transfer path is faster than the Hohmann ellipse. However, it requires a burn at A to reach an ellipse with a much higher apogee. At C, a combination of a horizontal and radial burns deflects it onto a circular path. This will consume more fuel than the equivalent Hohmann transfer.

up to the higher orbit it will gradually lose speed (as a ball thrown in the air does) until it reaches the apogee of the transfer orbit. At apogee, it is moving too slowly to follow a circular path (its motion is not circular since it is on an elliptical transfer orbit). At this point, another burn is needed to accelerate to the required speed.

It is also possible to transfer from a high orbit to a lower one using a Hohmann transfer with two burns to slow the craft down.

Of course, the Hohmann transfer is not the only orbit that can connect two others. It just happens to be the one that requires the least amount of propellant. It would be possible to burn the engines to achieve a greater ΔV, placing the spacecraft on an elliptical path with an apogee much higher than the target orbit (Figure 4.18). When the required orbit is reached, a combination horizontal and radial burn is required to stabilise the path. This process has the benefit of reducing transfer time at the cost of using more propellant.

In practice, burns are not impulsive, although high-thrust engines can be deployed to make the burn duration as short as possible (and so nearly impulsive). Hence more fuel is actually required than the idealised calculations suggest. Also, as orbits are generally elliptical rather than circular and are not arranged in a co-planar fashion, transfer orbits are not completely Hohmann in nature. With the ideal case (two circular co-planar orbits), the craft arrives at the second orbit 180 degrees from where it left the first. Any practical transfer orbit that covers less than 180 degrees is designated Type-1. Type-2 orbits cover more than 180 degrees. A lighter craft can use a Type-1 route; its smaller mass helps compensate for the greater mass of propellant needed. Heavier craft tend to take Type-2 routes. For reference, the *Mars Pathfinder* undertook a Type-2 flight and the *Mars Global Survey* utilised Type-1.

4.9 FLYING TO THE MOON

In 1989, a documentary film called *For All Mankind* was released. It featured commentaries by astronauts over original footage from their missions. The title of the film echoes the inscription on the plaque attached to one of the legs of the LM used in the Apollo 11 landing:

> Here men from the planet Earth first set foot on the Moon, July 1969 AD.
> We came in peace for all mankind.

The film contains a sequence shot from inside the command module out through one of its windows while the astronauts were onboard waiting for launch. The Moon is visible in a cloudless blue sky directly above the spacecraft. The astronaut commentary runs:

> I know they are doing their job right because the Moon is right straight ahead and that's where we are pointed, and they're just going to launch us right straight to this thing.

The business of flying a lunar mission is not as simple as pointing a rocket at the Moon and firing the motors. The Moon is a moving target, so the rocket must aim for where the Moon will be on arrival, not where it is at the time of launch. However, there are also some less obvious factors to consider, for example, the angle of sunlight on the lunar surface, to ensure good contrast and visibility for the landing.

After the systems were thoroughly checked out in Earth orbit, the Saturn V third stage fired again, placing the ship on a large elliptical orbit towards the Moon. Two possible mid-course corrections (to fine-tune the trajectory) using the service module's engine were budgeted for in the ΔV allowance.

As Apollo approached its destination, the Moon's gravity started to pull the craft towards it. If the service module engine failed to fire, the ship would loop around the Moon onto a free return to Earth (some later missions departed from the free return trajectory on the way out in order to explore other parts of the Moon's surface).

As Apollo was pulled behind the Moon, the service module engine fired to slow the craft into an elliptical orbit. On Apollo 11, the *lunar orbit insertion* (LOI) burn was a decelerating burn with a ΔV of -891 m/s which placed the craft into an elliptical orbit with an *apocynthion* (lunar equivalent of apogee) of 315 km and a *pericynthion* (lunar equivalent of perigee) of 111 km. Both measurements were taken to the surface of the Moon, not its centre. About four hours and two orbits later, the service module engine was fired again to circularise the orbit ($\Delta V = -48$ m/s) at 100 km by 120 km. This was a very well-designed orbit. The variations in the Moon's gravity (due to mascons and other factors) would make the orbit totally circular by the time the LM was due to redock with the command module after the landing.

The return trajectory was essentially the same as that on the outward trip. With the craft now considerably lighter (no LM) and the Moon's gravity less than that of the Earth, the service module engine was sufficient to accelerate the craft out of lunar orbit and onto an Earth-bound trajectory.

4.10 MISSIONS TO MARS

At first consideration, a journey to Mars is daunting. Apollo travelled at about 1.5 km/s on its way to the Moon.[11] At that speed, the trip to Mars would require 8.5 years of travel. No current technology is capable of building and equipping a spacecraft that could sustain life for that period of time.

However, the situation is not that bad. The average speed to the Moon is far less than might have been achieved with the Saturn V third stage. When properly fuelled, the stage could have boosted Apollo to 4.5 km/s, at which speed the lunar trip would have lasted just over a day. However, the service module engine would not have been powerful enough to brake Apollo into orbit about the Moon. The Moon's gravity helps the manoeuvre, but even so, the 1.5 km/s average used was about the highest practical speed.

A Mars mission does not have the same limitations. For example, Mars' gravity is considerably greater than that of the Moon, so it is capable of pulling in a faster object. Furthermore, unlike the Moon, Mars has a thin atmosphere that can help decelerate a spacecraft. The technique is called *aerobraking* and it was successfully used on both the *Mars Global Surveyor* and *Magellan* missions (the latter used radar to map the surface of Venus). An aerobraking manoeuvre dips the spacecraft into the upper layers of the atmosphere and uses the friction produced to help slow it down. Exploiting aerobraking means that less fuel needs to be carried and the craft can travel with far greater velocity.

Another factor may help to get a manned mission to Mars sooner than one might have thought. As the Earth moves around the Sun at ~30 km/s, the total speed of a spacecraft leaving Earth's orbit and heading for Mars is 30 km/s plus the launch ΔV. This is like throwing a ball from the front of a moving train. A craft launched with a ΔV of 3 km/s would move away from Earth's vicinity at a speed of 33 km/s relative to the Sun.[12] After various factors, such as the deceleration of the craft due to pulls from the Earth's and Sun's gravitational forces, are accounted for, we end up with a transfer time of ~250 days, which is far more manageable.

This extra boost does not significantly affect the braking into Martian orbit at the other end. Mars moves around the Sun, at 24 km/s and the spacecraft will be nearly parallel to the planet on arrival. That means a relative velocity of 3 km/s since the craft by then will have slowed to ∼21 km/s.

After all these factors are taken into account, we reach the surprising conclusion that a mission to Mars can be flown from low Earth orbit with a total ΔV that is actually *less* than that required to travel to the surface of the Moon (see Figure 3.9).

4.10.1 MARTIAN TRANSFER ORBITS

Mission designers have considered a wide range of possible transfer orbits to Mars. They fall into two broad classifications: *opposition-type* orbits and *conjunction-type* orbits.

Consider Figure 4.19, which is an outline of the Sun–Earth–Mars system. As Mars takes approximately 2 years to circle the Sun, the planet at times is considerably closer to Earth than it is at other stages of its orbit. From the perspective of an observer on Earth, when Mars is in position M1, it appears in a portion of the sky opposite to the Sun. Hence, this planetary alignment is called *opposition*. Confusingly, when the planets are close they are in opposition.

On the other hand, if Mars is at M2, it can be viewed from Earth in the same region of the sky as the Sun; this is called *conjunction*.

A conjunction-type trajectory arrives at Mars when the planet is in conjunction relative to the Earth's position at the time of departure. An opposition trajectory will have Earth and Mars on the same side of the Sun at some stage of the journey.

The selection of an outbound trajectory must consider the location of Mars in its orbit when the crew arrives. The same is true in reverse for the return leg, but unfortunately, during the trip out, the Earth will have moved in its orbit and will be in a relatively unfavourable position (from a ΔV view) for an immediate return. Clearly, some surface stay is needed, but the stay will have an impact on the alignment and must be optimised as part of the mission design. As a result, a combination of factors dictates optimum transfer selection: launch date; total ΔV available (or desired); transfer duration and surface stay duration.

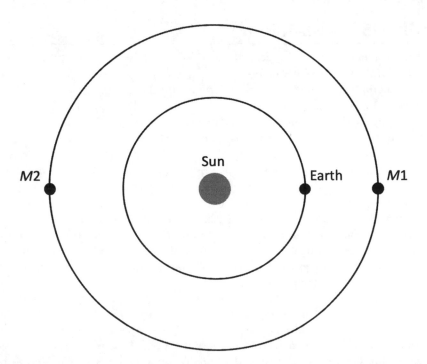

Figure 4.19 The relative positions of Earth, the Sun and Mars when Earth and Mars are in opposition (M1) and conjunction (M2).

Opposition-type transfers

These missions tend to have one short duration transfer leg and one longer duration transfer. They can be in either order: the short leg can be scheduled on the way out or on the way back. However, the long leg can benefit from an unusual trajectory that swings past Venus and uses the planet's gravity to accelerate (on the way out) or decelerate (on the way back) the craft without needing fuel to achieve the ΔV (Figure 4.20). The choice is dictated in part by the alignment of Earth and Venus, but other considerations play a part.

If the mission designers elect to schedule the long transfer at the start of the mission, the crew have to contend with ~360 days of zero-g before recovering on the Martian (0.38 g) surface. This raises some concerns about their ability to cope with the accelerative manoeuvres associated with orbital insertion and landing. Hence the long leg is generally placed in the return phase.

Total mission times vary, depending on launch dates but range from 560 to 850 days (1.5 to 2.4 years). Surface operations fall between 30 and 60 days. Generally, missions of this class have high overall ΔV requirements due to their sub-optimum transfers. They also exhibit wide variations in ΔV depending on the exact dates chosen for departure and return (see Figure 4.22 and the discussion below). Employing a Venus flyby presents challenges for thermal and radiation shielding.

Conjunction-type transfers

These missions have the global minimum ΔV requirement for any given launch date. Instead of departing Mars on a non-optimal return trajectory, the surface stay is extended, while the mission awaits a more favourable alignment of the planets (Figure 4.21). The need for optimum conditions can mean up to 550 days on the surface and total mission durations around 1,000 days (2.7 years). Another advantage of this class of mission is that the ΔV requirements show only minimal variation based on the opportunities for launch and return. However, the longer duration of the mission as a whole presents challenges for the transport of consumable supplies, prevention of radiation exposure and equipment durability.

ΔV versus mission duration

Figure 4.22 shows the results of extensive calculations performed for NASA's Design Reference Architecture 5.0 report.[13] The far right of the graph contains a single descending line from a ΔV of 30 to ~6 km/s (marked with diamond-shaped data points). This curve corresponds to long-stay

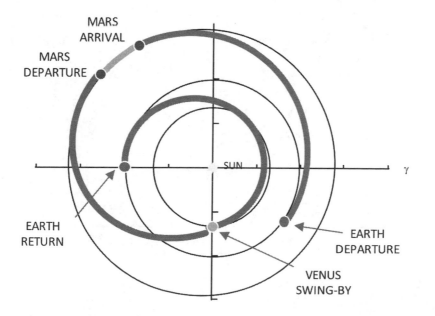

Figure 4.20 An example opposition (short-stay) mission profile employing a Venus flyby on the return leg. (Image courtesy of NASA. Figure 2.1a from NASA/SP–2009-566-ADD2.)

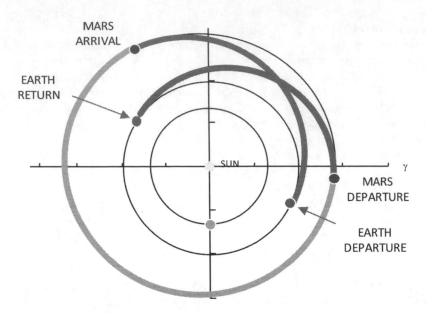

Figure 4.21 An example conjunction (long-stay) mission profile. (Image courtesy of NASA. Figure 2.1b from NASA/SP–2009-566-ADD2.)

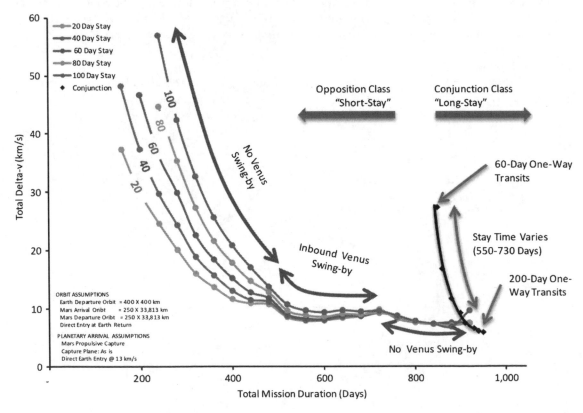

Figure 4.22 ΔV trade-off against mission duration for opposition and conjunction mission profiles. Various assumptions have gone into the calculations for departure year 2033. (Image courtesy of NASA. Figure 2.2 from NASA/SP–2009-566-ADD2.)

Orbits and trajectories

(conjunction-type) missions and shows how the mission duration varies with the ΔV allocated. In these scenarios, mission time is largely driven by the duration of the surface stay, which must be longer in order to wait for a low-energy return. Sufficient total ΔV can decrease transit times down to 120 days but the total duration remains ~900 days.

The curves across the left side of the graph apply to short-stay (opposition-type) missions. Data for Venus swing-by options are included as well. The curves represent various periods spent on the surface (20 to 100 days). Several points are clear:

- The ΔV requirements increase dramatically as mission time is reduced, especially below 500 days without a Venus swing-by.
- Shorter mission durations are more sensitive in ΔV terms to surface stays. A 300-day mission profile needs a ΔV of ~19 km/s for 20 days on the surface (7% of the mission) and a ΔV of ~39 km/s for a 100-day stay (33% of the mission).
- With mission durations exceeding ~900 days, the ΔV is a little more than that needed for a conjunction-type mission of the same duration.

Further considerations can impact transfer times if continuous low-thrust propulsion or in-orbit refuelling is employed.

4.10.2 MARS LAUNCH WINDOWS

Every 780 days, Mars and Earth draw closer together due to the relative periods of their orbits. However, given Mars' rather elliptical orbit, some oppositions are closer than others; the best ones occur when Mars is closest to the Sun. They take place on a roughly 15-year cycle and need only half the ΔV of the other oppositions (Figure 4.23). Over the range of dates shown, the closest opposition will be in 2050 with a

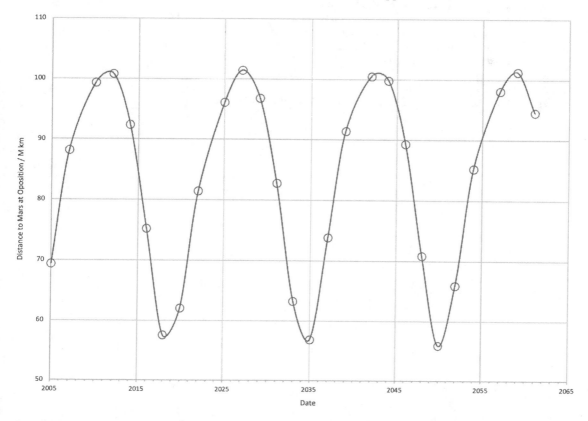

Figure 4.23 The distance to Mars at opposition. The approximate 15-year cycle is clearly visible, as are the excellent launch opportunities coming up in 2033 and 2035. (From http://spider.seds.org/spider/Mars/marsopps. html.)

distance of 55.96 million kilometres and the furthest in 2027 at 101.42 million kilometres – clearly a very large difference. This explains why the 2030s are seen as crucial opportunities for human explorations of Mars. Chapter 9 will present a somewhat deeper look into the plans and options for Mars missions.

ENDNOTES

1. This simulation of gravity will be used in large space stations but is different from what science fiction authors think of as artificial gravity: producing a gravity field without the necessary mass.
2. The craft did spin slowly in what was known as the *barbecue roll* to ensure that all areas of the spacecraft were exposed equally to the heat from the Sun and the cold of space. If the craft maintained the same attitude permanently, the area facing the Sun would have baked.
3. Terminal velocity is not so named because of the fatal result from hitting the ground. It is the velocity reached as the forces on a body become balanced when acceleration stops.
4. The only situation that comes close occurred during the Apollo flights when the spacecraft momentarily passed the point between the Earth and the Moon where the two forces of gravity exactly balanced. However, even then, the Sun's gravity would have pulled on the astronauts.
5. Douglas Adams, in *The Hitchhiker's Guide to the Galaxy,* refers to flying as 'learning how to throw yourself at the ground and miss'. That statement certainly applies to being in orbit.
6. Of course, the required centripetal force also decreases as we make the radius of the circle larger. However, the centripetal force decreases like $1/r$. Gravity decreases by $1/r^2$, so the gravity weakens more rapidly with distance than centripetal force does. To compensate, the speed must decrease as well.
7. Technically speaking, the boots do not generate the centripetal force. Recalling our discussion of Newton's third law of motion in Chapter 3, it is clear that the magnetic boots exert a force on the metal walls of the drum. The reaction force of the drum on the boots and hence on the astronaut produces the centripetal force on the astronaut.
8. Strictly speaking, the centre of gravity of the combination is at one focus. However, in our solar system, the Sun is far more massive than all the planets put together. Thus the centre of gravity of our solar system is virtually in the same place as the centre of gravity of the sun.
9. The tangent line at any point on a circle is a line drawn at 90 degrees to the radius at that point on the circle.
10. This motion is genuinely horizontal only in the sense that it is parallel to the ground at apogee and perigee.
11. This is an approximate average speed. The spacecraft would have started moving more rapidly and be slowed by Earth's gravity. As it moved into the region where the Moon's gravity started to dominate, the craft would start to accelerate again.
12. This effect is no help on a Moon mission because the Moon travels with the Earth.
13. https://www.nasa.gov/sites/default/files/files/NASA-SP-2009-566-ADD2.pdf

Intermission (2)

From Mercury to Gemini

Having successfully placed the first American satellite in orbit, van Braun's team were cut loose and given the brief to get 'man in space soonest'; the Mercury space programme was born. NASA set about selecting seven men to form a new *astronaut corps* and engineers began modifying the Redstone and Atlas missiles. Redstone was not powerful enough to accelerate a manned capsule up to orbital speed, but it was capable of lobbing a man on a ballistic flight out of the atmosphere. This would give NASA valuable experience of maneovring in space and re-entry. In the meantime, the far more powerful Atlas was readied for orbital missions.

Compared to the Gemini and Apollo craft that followed, the Mercury capsule was tiny – just under 3 m tall and 1.9 m wide at the base, bell shaped, and housing a pressurised compartment barely large enough to hold one man. Mercury was the sort of ship that an astronaut wore rather than flew (Figure I2.1).

In 1961, Cape Canaveral was the scene of frantic activity. Test flights of the Redstone and Atlas rockets carrying unmanned Mercury capsules proceeded with varying degrees of success. The Redstone proved to be quite reliable; the Atlas was another matter. Its hull material was so thin in places that only its internal pressures kept it from collapsing. Atlas also had an alarming tendency to fail during flight, once its warhead was replaced by a Mercury capsule.

The climate of competition was intense. Soviet representatives[1] never missed an opportunity to comment publicly about the relative achievements of the USSR and US space programmes. NASA moved determinedly but cautiously. Alan Shepard was selected as the first Mercury 7 astronaut to ride a Redstone into space but he had to wait for his turn. After launching successful unmanned flights, the next step involved a chimpanzee named *Ham*. While the astronauts trained in simulators, several chimps were taught to bang levers in space. They were rewarded by banana pellets and punished by jolts of electricity. Ham was selected as the best of his team.

The highly qualified, supremely fit and ambitious Mercury 7 astronauts did not take kindly to riding second to a chimp, but the doctors felt that space flight involved too many unknowns. With hindsight, it is hard to appreciate the level of uncertainty: would zero-*g* distort the eyeball and prevent astronauts seeing properly, for example?

Despite Alan Shepard's vigorous campaign to have the chimp grounded, the doctors prevailed. On January 31, 1961, the astronauts gathered at Cape Canaveral to watch Ham take his flight. He performed flawlessly; the rocket did not. The Redstone burned its fuel too quickly. The escape tower system sensed the fuel problem and fired to lift the capsule clear of the rocket. The acceleration was brutal. In the cabin, Ham proceeded to hammer every lever that he was trained to hit, but the system greeted him with shocks instead

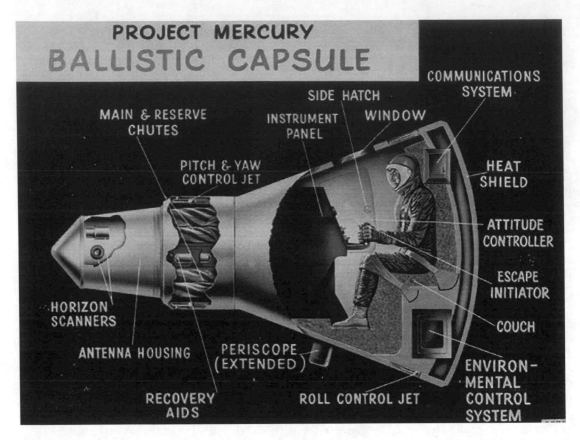

Figure I2.1 The tiny Mercury capsule. (Image courtesy of NASA.)

of rewards no matter what he did. The capsule sailed nearly 200 km further from Cape Canaveral than anticipated and crashed into the water. By the time recovery helicopters arrived, the capsule was half full of water and Ham nearly drowned.

The fault in the Redstone was traced to an electrical relay. Shepard pressed to take the next flight but von Braun stepped in. Feeling the responsibility of a man's life, von Braun wanted another test flight. That test flew perfectly on March 24, 1961. Unfortunately, Yuri Gagarin made one complete orbit of the Earth on April 12, 1961 while Shepard's Redstone sat on the pad having final adjustments. The propaganda coup was crushing. In the middle of the night, as Gagarin passed over America, reporters were looking for some comment from NASA. One official roused from sleep by a reporter unguardedly replied: 'If you want anything from us, you jerk, the answer is we're all asleep.' The next morning's headlines read:

SOVIETS SEND MAN INTO SPACE;

SPOKESMAN SAYS U.S. ASLEEP.

Two days later, President Kennedy called his advisors to his office and asked how the United States could catch up. He tasked Vice President Lyndon B Johnson with finding a way to beat the Soviets. In the meantime, the American programme seemed to fall apart. Another Atlas failed and a smaller rocket designed to test the Mercury escape system crashed. Top people in the White House urged the president to cancel the manned programme as it was too dangerous. Considering the unfavourable publicity to date, the advisors asked the White House to imagine what would happen if an American died on a NASA rocket. Johnson urged the president to continue. The failures did not involve Redstone rockets and NASA needed time to correct the Atlas problems. Shepard's flight on May 5, 1961 lasted 15 minutes and 22 seconds (Figure I2.2). Twenty days later, Kennedy announced to Congress that the United States would go to the Moon.

Figure I2.2 The launch of Alan Shepard's sub-orbital flight. (Image courtesy of NASA.)

The Mercury programme continued. The second flight was another sub-orbital hop on a Redstone. It was a great success until shortly after splashdown when the hatch mysteriously blew open filling the capsule with water. Astronaut Gus Grissom (later to die tragically in the Apollo 1 fire) scrambled out as the craft sank.

The first Atlas flight carried John Glenn into orbit and an adventure that made him an American hero. During his 4 hour, 55 minute, 23 seconds flight, he orbited the Earth three times in the tiny *Friendship 7* capsule.[2] As he passed over Western Australia, all the lights in the city of Perth were turned on to welcome him. Unbeknownst to Glenn, a fault indicator at mission control suggested that his heat shield might have come loose. As a result, the ground controllers asked him not to jettison the retro-rocket pack that would normally be released after the rockets fired to slow the craft into its re-entry corridor. They were worried that the pack might be the only thing holding the heat shield in place.

Glenn knew that something was seriously wrong. Keeping the retro-rocket pack in place was a drastic change of mission plans. As a result, the autopilot that should have guided the capsule during re-entry had to be over-ruled and Glenn flew the dangerous passage manually. He later admitted to a few anxious moments watching pieces of metal fly past the window. Fortunately, it was only the retro-pack burning up not the heat shield failing. If the mishap had occurred on a Soviet flight, no one would have known about it; Glenn's recovery of the mission captivated the world.

Kennedy was reluctant to lose a genuine American hero to the dangers of space travel and grounded Glenn after only one flight. However, in 1998, at the age of 77, Glenn finally flew again onboard the space shuttle *Discovery* as part of the STS95 mission. While many regarded the flight as a publicity stunt, it yielded some scientific data. Tests performed on Glenn explored similarities between the physiological effects of space flight and aging. Data from STS95 was compared to data from Glenn's original Mercury mission.

NASA-S-65-893

Figure I2.3 The Gemini capsule. (Image courtesy of NASA.)

The next three Mercury flights culminated in Gordon Cooper's 22 orbits on May 15, 1963. This flight also ended with electrical problems onboard and required Cooper to fly the craft through re-entry manually. He splashed down only 4 miles from the recovery carrier.

Although this flight stretched the Mercury capsule to its limits, the United States did not seem to be catching up to the USSR. On June 14 1963, the Soviets put a man into orbit for a staggering 119 hours, and 2 days later it sent the first woman into space. On the American side, things went rather quiet.

The decision to go to the Moon kick-started a debate within NASA as to how best to achieve such a goal. The Apollo programme had congressional funding and von Braun's team continued its work on booster configurations. However, the exact mission profile had not been decided and this was impacting the design process. The matter was settled in 1962. Despite early reservations, NASA chose a mission that required two spacecraft to find one another in orbit and dock while in orbit around the Moon. Clearly NASA and the astronauts (now 15 in number) had to become experts in rendezvous and docking and quickly. This was the philosophy behind the Gemini programme.

Gemini was a much larger two-man capsule designed to be highly manoeuvrable in space (Figure I2.3). The Mercury programme had shown that a capsule could be steered along its path, but Gemini was designed to change orbits and dock with other craft.

In October 1964, the Soviets flew the first three-man mission. On May 18 1965, Aleksey Leonov performed the first spacewalk, 5 days before the first Gemini flight, the date of which was announced in February.

The Americans had sent unmanned probes to study the lunar surface is preparation for Apollo, but to the public it looked like nothing significant happened between the last Mercury flight (May 1963) and the first Gemini test (April 1964). In the midst of this hiatus, Kennedy was assassinated. The programme lost the man responsible for setting it in motion. Fortunately, President Johnson was an even more passionate supporter of the space programme.

The first manned Gemini flight took place on March 23, 1964. Onboard were Gus Grissom, the first man to fly in space twice, and John Young. Over the next 18 months, nine more Gemini flights pushed

Figure 12.4 Gemini 7. (Image courtesy of NASA.)

NASA to the front of the space race. In that time the astronauts perfected the techniques of rendezvousing with robot target craft and with other manned Gemini capsules (Figure 12.4).

Space walking was refined, although not without difficulty. In case of a docking failure, it was vital that the Apollo crew could transfer from the lunar module to the command module by a spacewalk. Experiences on Gemini missions showed that it was very hard to grip the smooth sides of spacecraft and also how simple tasks on Earth required considerable time and effort in space.

Gemini was not without drama. Gemini 8 carrying David Scott and Neil Armstrong nearly ended in disaster when one of the manoeuvring thrusters jammed open, putting the craft into a spin so violent that the crew nearly blacked out. Only very quick thinking by Armstrong saved the mission. He used the re-entry thrusters to stabilise the ship while the jammed thruster used up its fuel.

The last Gemini flight was a masterpiece. Shortly after launch, Jim Lovell (later to command the ill-fated Apollo 13 and already a veteran of a previous Gemini flight) and Buzz Aldrin (later to land on the Moon in Apollo 11) discovered that their onboard computer system did not function properly. Fortunately, Aldrin (the only member of the astronaut corps with a PhD) had done much of the theoretical work on the manoeuvres needed to rendezvous with another craft in orbit. He was NASA's acknowledged expert on such matters. Using a sextant and charts, he guided the Gemini to the designated robot craft. Later, his intense study of the problems involved in spacewalking and the improvised tools he carried enabled him to carry out a series of complex tasks outside the craft.

The scene was set for the Apollo programme to follow. However, NASA was about to face its first real tragedy. It would not be the loss of life on a mission as NASA and the test pilot astronauts had always known would be a possibility. America would lose its first members of the astronaut corps during a fire in the Apollo I capsule on the ground during a routine test.

ENDNOTES

1. While visiting the set of the film *2001: A Space Odyssey* in May 1966, the Soviet Air Attaché commented that all the labels on the controls of the American spaceship *Discovery* featured in the film should have been in Russian.
2. The names of all Mercury capsules ended with the numeral 7 in honour of the first seven astronauts chosen to lead the American space programme.

5 The Apollo command and service modules

5.1 MISSION MODES

When US President John F Kennedy issued his challenge in May 1961, NASA's long-term strategy planners had already been considering various ways of landing a man on the Moon. Most of them decided that a direct flight would involve the least risk to the crew. Such a mission was to start with the launch of a huge rocket, the upper stage of which would fly directly to the Moon. The craft might go into orbit temporarily then land or possibly proceed directly to the surface. Having completed their exploration, the astronauts would then fire the engines again and use the ship to return to Earth.

Proponents of the direct flight idea rejected the alternatives as most of them involved docking two spacecraft. At the time, NASA had no experience of manoeuvring ships with the precision necessary for docking. The Mercury astronauts had shown that control of a capsule in space was possible, but docking would require very fine adjustments of speed and direction, as well as the ability to alter orbit. It was not certain that thrusters could provide such flexibility.[1] Even if such manoeuvres could be performed, the project still faced the complex orbital mechanics involved in getting the two craft close to each other for docking.

However, direct flight had its drawbacks as well. The landing craft would have to carry to the lunar surface sufficient fuel for the return journey to Earth.[2] Adding to the weight would be the need for a heat shield for surviving re-entry and the storage of consumables sufficient for all of the astronauts for the whole voyage. The craft would end up being extremely heavy, placing severe demands on the landing legs and the main engines. Furthermore, the booster rocket needed to launch the mission from the surface of the Earth would be much larger than any that NASA had ever developed.

The only significant competition to direct flight came from Wernher von Braun's missile group. His team favoured an *Earth Orbit Rendezvous* (EOR) mission that would involve launching several boosters carrying parts and fuel for a Moonship. After the craft was assembled and fuelled in Earth orbit, it could depart for the Moon. From that point, the mission would proceed in exactly the same manner as the direct flight. Although EOR would require several launches to lift all the fuel and components into orbit, von Braun's team thought that this proposal was technically preferable to trying to develop the massive booster required for the direct flight.

After transferring from the US Army to NASA in 1960 and effectively setting up the Marshall Space Flight Center, von Braun's team continued to work on the EOR mission profile.

The Marshall Space Flight Center was not the only NASA facility with ideas on Moon exploration. As early as 1959, the Jet Propulsion Laboratory (JPL) in Pasadena, California, championed *Lunar Surface Rendezvous*. The goal was to send a number of unmanned rockets to the Moon in advance of the astronauts' flights. One rocket would be an Earth return craft and one or more fuel tankers would accompany it (Figure 5.1). A ground control team would then refuel the Earth return vehicle automatically and confirm that it was fit to fly again, before sending the astronauts into space. This plan was considered too risky by senior management at NASA, but it is interesting to note that a similar idea has been considered for a future Mars mission (see Chapter 9).

Exercising the minds of all the mission planners was America's lack of a large payload booster in the early 1960s. The first American in space had been launched by the Mercury Redstone (a modified ballistic missile). The Atlas and Titan II boosters developed for later Mercury and Gemini missions were not powerful enough for Apollo. Plans for two new boosters were underway: *Saturn* and the even larger *Nova* family. Saturn was envisaged as a family of boosters capable of launching Earth orbit and lunar orbital

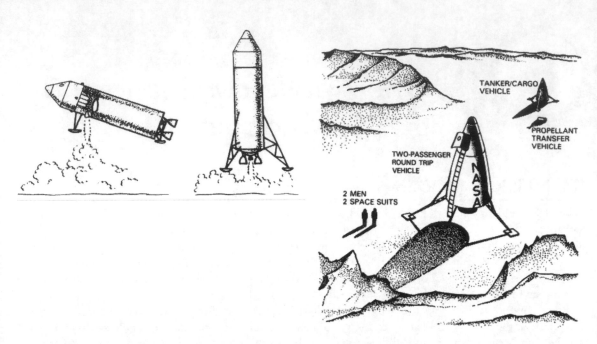

Figure 5.1 Ideas for sending a man to the Moon. Right: Lunar surface rendezvous. A return vehicle and a tanker for refuelling are sent to the Moon ahead of the crew. Left: Two possible landing modes for a giant rocket suitable for direct ascent. (Image courtesy of NASA.)

flights. Von Braun's EOR flight plan involved several Saturn flights. The Nova family of boosters were being developed in parallel to provide the capability for hurling a spacecraft to the Moon on a direct flight. The most representative design, often used in comparisons, was the Nova C-8. This vast booster required eight F-1 engines in its first stage, would have stood 3 m taller than the Saturn V and measured 5 m wider in diameter at the base of the first stage. There was even some thought of using nuclear propulsion in the upper stage.

When President Kennedy announced his challenge before the world, the F-1 engine and the Saturn needed a great deal of development. Decisions about how the booster programme should continue were affected by the mission profile discussion, which was in turn influenced by the booster plans. The need for a commitment to one of the possible modes became urgent.

5.1.1 THE WILD SIDE

With the pressure to land a man on the Moon as soon as possible, some wild ideas started to surface. Every proposal was considered, from refuelling a spacecraft en route to the Moon to landing a man on the surface and keeping him alive with regular supply rockets until NASA could find a way of getting him home!

One idea sparked in this period was destined to be rather more significant: *Lunar Orbit Rendezvous* (LOR). In this scheme, a lunar lander would detach from a spacecraft in orbit around the Moon and carry astronauts to the surface. At departure time, the lander would take off and dock with the command module that remained in orbit. When this idea was proposed in 1959, it was seen as a way of reducing the total weight of the spacecraft as components could be abandoned at appropriate moments in the mission after they fulfilled their purpose. Effectively this scheme carried the principles of staging from booster design to its logical conclusion.

Historians of NASA during this period generally credit Thomas Dolan of the Vought Astronautics Division with carrying out the first design study of the LOR mission mode. He assembled a team to analyse lunar missions and ways in which his company might contribute to NASA's plans. Vought was not awarded any of the contracts for Apollo, but the company's results turned out to be prophetic. Dolan's team fully understood that ΔV budgets are the key to mission design. They proposed a modular spacecraft, with separate components performing different functions. Dolan concluded that the best approach was to discard pieces that were no longer needed. Furthermore, he saw no reason to take an entire spacecraft down to the lunar surface and back to lunar escape velocity. Dolan's team did this work in 1959.[3]

Unbeknown by Dolan and other NASA staff at the time, LOR had been studied by Yuri Kondratyuk,[4] a Ukranian engineer, in 1916 and then by H E Ross, a British scientist, in 1948.[5] Inherently correct solutions to engineering problems have a tendency to occur in a variety of contexts.

John Houbolt of NASA's Langley Research Center was working on rendezvous issues when his calculations suggested LOR as an effective weight-saving technique. These initial ideas from early 1960 grew ever stronger until in the summer of that year, some further calculations convinced Houbolt 'that lunar-orbit rendezvous offered a chain reaction simplification on all "back effects"; development, testing, manufacturing, erection, count-down, flight operations, etc.' As he wrote in a later article,[6] 'I vowed to dedicate myself to the task.'

With hindsight, it seems curious that NASA took so long to adopt LOR, given the technical advantages inherent to this mode. Houbolt and others pressed the idea on the basis of weight saving, and the consequent simplification of the landing as well as the resulting reduction in booster power necessary to launch the mission. Time pressure exerted on booster development steadily deflected thinking away from the huge Nova rocket that would have been required for a direct ascent, swinging thinking towards EOR and laying some of the groundwork for LOR acceptance.

As designs for Apollo progressed, the key advantage of being able to tailor make various modules of the spacecraft to achieve different jobs started to become apparent: the command and service modules did not need to be compromised for lunar landing and the lunar module did not have to survive re-entry. However, other engineers within NASA felt that the impressive weight reduction figures quoted for LOR were overly optimistic, given the extra complexity in systems and sub-systems needed for a modular craft (life support in two independent modules, for example, as well as multiple engine systems, guidance equipment and other components).

Over-arching these technical debates was a powerful emotional issue as well. Should some part of a LOR mission fail, NASA would face the prospect of having dead astronauts in orbit around the Moon. If the worst happened in Earth orbit, easier possibilities for retrieval or at least decaying an orbit to burn up would be available. Timing was also clearly a factor. Initial proposals for LOR were in progress before Kennedy tasked NASA with landing on the Moon. At the time, management was still working through the Mercury missions, with little mental energy left for considering future phases that involved risky manoeuvres.

A historical study of the path to the final decision is convoluted and various analysts have reached slightly different conclusions. Such differences are to be expected given the many technical, political and emotional factors involved. Houbolt's perseverance and energy were almost certainly critical in breaking through the reluctance to consider the LOR option. Perhaps significant was Houbolt's expertise in the theoretical aspects of rendezvous in general. He successfully docked craft in a simulator in 1959 and often presented LOR as a possibility during his presentations on rendezvous in general.

A key moment in NASA's conversion to LOR came when von Braun turned to the idea after having been one of the most vigorous proponents of EOR. He announced his change of mind in June 1962 at a meeting convened to present the results of the Marshall Space Flight Center's continuing studies of EOR. After listening to members of his team describe their EOR plans, von Braun spoke from notes made during the meeting. To his astonished audience, he announced that the centre would support LOR. He explained the switch in the following way:

I would like to reiterate once more that *it is absolutely mandatory that we arrive at a definite mode decision within the next few weeks....* If we do not make a clear-cut decision on the mode very soon, our chances of accomplishing the first lunar expedition in this decade will fade away rapidly.

He continued to say that LOR 'offers the highest confidence factor of successful accomplishment within this decade'. The key reason being the separation between the design of the landing craft and the command module: 'A drastic separation of these two functions into two separate elements is bound to greatly simplify the development of the spacecraft system [and] result in a very substantial saving of time.' From that point, all the NASA facilities adopted LOR as the way to get a man to the Moon.

5.2 THE COMMAND MODULE (CM)

The contract for the design and manufacture of the Apollo command module was granted before the decision to go with LOR was made. By late 1962, plans were well underway for the command module of an Earth orbital and circumlunar craft (Figure 5.2). The contractor, North American, had given some thought to attaching a propulsion stage for a landing on the Moon. However, this need was circumvented by the switch to LOR.

The greatest impact that the decision to go with LOR had on the design of the command module was the need to incorporate a docking mechanism with the ability to transfer two astronauts to the lunar module (Figure 5.2). Consequently, the development of the command module continued along two parallel courses. The early design, not including docking facilities, became the Block I spacecraft used in astronaut training and was destined for Earth orbital missions. However, due to the pressure of time imposed by the presidential deadline, North American could not wait for the Block I ship to be tested and flown before work on Block II (docking version) could begin. In fact, the Block I design never flew a manned mission.[7]

Clearly, some components were common to the two versions and North American subcontracted for systems such as the escape tower with its solid rockets (Figure 5.3), the command module manoeuvring thrusters and the heat shield.

The basic design parameters for the command module called for it to function as a 'combination cockpit, office, laboratory, radio station, kitchen, bedroom, bathroom, and den'. The specification also

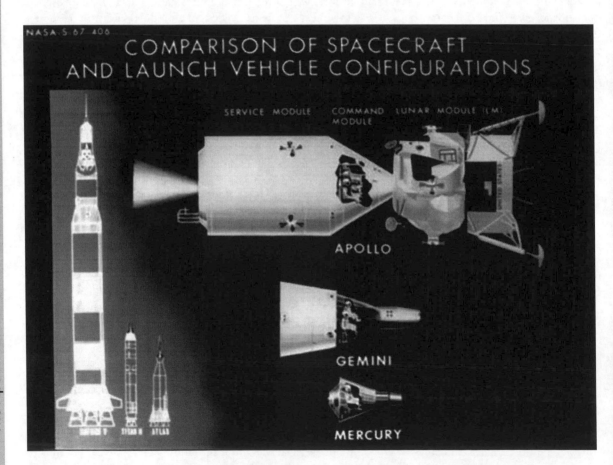

Figure 5.2 A NASA graphic illustrating huge advances in spacecraft design implied by Apollo. Left: Saturn, Titan II and Atlas boosters are compared. Right: Comparison between the Mercury, Gemini and Apollo spacecraft. (Image courtesy of NASA.)

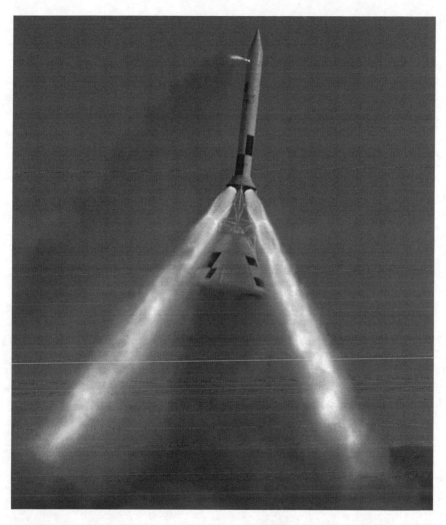

Figure 5.3 Testing the launch escape system designed to pull the whole command module clear of the Saturn V should a problem develop during launch. The command module would then use its parachutes to land. (Image courtesy of NASA and Mike Dorffler.)

mandated a shirtsleeve environment within the capsule, as it was unrealistic to expect the astronauts to remain in their spacesuits for the duration of the journey.

Designers rapidly settled on the Apollo command module as a refinement of the Mercury capsule used at the start of the manned space programme (Figure 5.2). The blunt-ended cone shape had proven to be effective in surviving the rigours of re-entry. Calculations showed that the Apollo capsule should be more tapered than Mercury because the higher entry speed experienced by a ship returning from the Moon would generate a shock wave in the atmosphere. This, in turn, would cause a super-hot region to form around the capsule, so hot that the heating effect of the infrared radiation produced would significantly raise the external temperature of the craft. This extra heating mechanism (in addition to conduction and convection in the atmosphere) was not experienced by the slower entering Mercury capsules. The more tapered shape helped to mitigate this effect.

The choice of splashdown into the ocean rather than a land-based return was forced by the geographic position of the launch site. Cape Kennedy's location on a peninsula (for safety reasons, large potentially dangerous bombs [rockets] are not launched near major population centres) meant that if an abort took place during the launch the capsule would have to land in the water anyway, so it made sense to adopt that as the preferred landing option.

5.2.1 COMMAND MODULE CONSTRUCTION AND HEAT SHIELDS

Height: 3.22 m	
Diameter: 3.91 m	
Mass (including crew): 5,900 kg	
Mass (splashdown): 5,310 kg	

The command module subdivided into three compartments (Figure 5.4):

- forward compartment – the small space at the point of the cone;
- crew compartment – within the primary structure at the centre of the capsule;
- aft compartment – another small space around the outside edge of the lower half of the primary structure.

The primary structure was constructed from a welded aluminium inner skin to which was glued an aluminium honeycomb[8] faced with additional sheet aluminium. The honeycomb varied in thickness from 6.4 cm at the top (near the docking structure) to 3.8 cm near the aft heat shield. On top of the primary structure sat a cylinder that reached to the very summit of the capsule. This was the tunnel through which the mission commander and lunar module pilot would enter the lunar module. Prior to docking, the probe mechanism was located at the end of this tunnel.

Forming the sloping sides of the cone was the outer structure manufactured from stainless steel brazed into a honeycomb and in turn brazed to the steel alloy inner and outer sheets. The outer structure formed one of the principal barriers to protect the astronauts and their equipment from the extreme temperature variations expected during the mission. Consequently, a heat shield was bonded to the cone's outer surface.

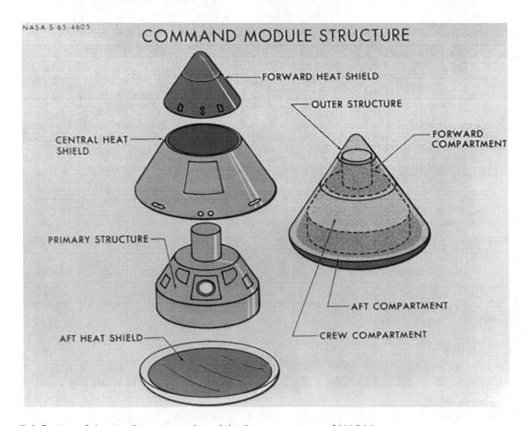

Figure 5.4 Design of the Apollo command module. (Image courtesy of NASA.)

The Apollo command and service modules

During liftoff, friction with the atmosphere raised the external temperature to 650°C. Once in space, the side of the spacecraft facing towards the Sun would be baked by sunlight to 140°C, and the side away from the Sun would freeze to −170°C. Finally, during re-entry, the external temperature would rise to, 2800°C, although the brunt of this would be taken by the aft heat shield.

In flight, the spacecraft was oriented at 90 degrees to the direction of travel (something that films often get wrong) and set into a slow rotation (known colloquially as the *barbecue roll*) which ensured that each side received an even roasting from the Sun (Figure 5.5).

The heat of the launch was dealt with by the boost protective cover. This was another cone, made from fibreglass with a white painted cork outer cover, that fitted over the command module and was fixed to the launch escape tower. Fortunately, the launch escape system never had to be used during a mission. It was ejected along with the boost cover at about 90 km altitude.

Bonded onto the outer structure was the main heat shield (Figures 5.4 and 5.6). This consisted of 400,000 plastic honeycomb cells hand filled with an ablative material designed to protect the crew from extreme temperatures. Such materials char and melt rapidly, absorbing and dissipating energy as they do so. At the time, no available material could fully prevent thermal conduction at such high temperatures.[9] By progressively melting and falling off, an ablative heat shield can form a very effective protective barrier.

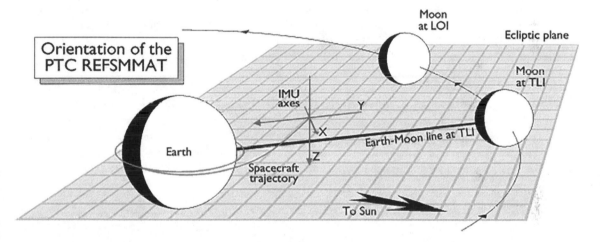

Figure 5.5 Spacecraft orientation during Earth–Moon flight (from the Apollo 15 flight transcript). The z-axis is along the line from the service module engine bell to the lunar module descent stage engine bell. The stack was set into rotation about this axis during flight so that all sides would be evenly warmed by sunlight. The PTC REGSMMAT caption refers to the guidance platform's reference orientation during the Passive Thermal Control manoeuvre (barbeque roll). (Image courtesy of NASA.)

Figure 5.6 Stages in constructing a command module. (Images courtesy of NASA and North American.)

The aft heat shield was constructed in the same manner as the sides of the cone, but was somewhat thicker (up to 5 cm). The final outer covering over all the heat shields was a moisture barrier and a Mylar thermal coating that looked like aluminium foil.

The completed command module was as light as possible, yet rugged enough to survive the heat of re-entry, the impact of splashdown and possible collisions with small meteorites.

5.2.2 THE FORWARD COMPARTMENT

This section of the CM formed a ring around the docking tunnel at the top of the primary structure. It was divided into four equally sized segments containing all the equipment necessary for landing (parachutes, beacon light, recovery antenna, flotation balloons), two small manoeuvring motors and the release mechanisms for the forward heat shield (the shield had to be ejected before the parachutes could be released). The forward heat shield had four slots for the legs of the launch escape tower, which were attached with bolts that contained small explosive charges. When the time came to jettison the launch escape tower and the boost cover, the charges would fire and break the bolts.

5.2.3 THE AFT COMPARTMENT

Running around the outside of the lower part of the primary structure and above the aft heat shield, the aft compartment was divided into 24 bays. These bays contained 10 more manoeuvring engines with the fuel and oxidiser tanks for the whole command module manoeuvring system, water tanks, ribs designed to crush on impact with water (just like the deforming structure in a car that absorbs energy in a crash), some instruments and the junction where wiring and plumbing ran from the command module to the service module.

5.2.4 THE CREW COMPARTMENT

The crew compartment inside the primary structure contained the living and working environments for three men during the majority of the mission. Inside was a habitable volume of just under 6 m³ (Figure 5.7).

The cabin was air conditioned and the temperature regulated at 21 to 24°C. A pure oxygen atmosphere was used with a pressure of 35 kPa in flight.[10] Exhaled gases were recycled and the CO_2 contaminants removed so that the remaining oxygen content could be returned to the atmosphere. The environment unit

The Apollo command and service modules

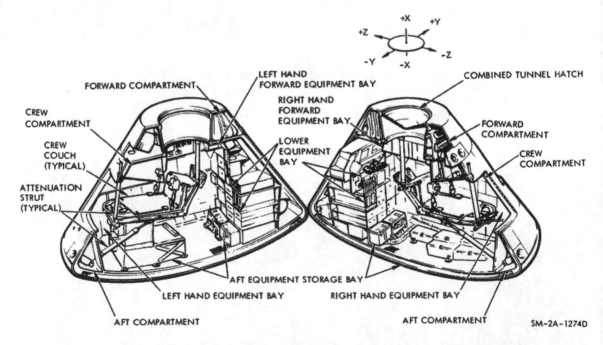

Figure 5.7 The interior of the command module (centre couch not shown) showing the various equipment bays. (From *NASA Apollo Operations Handbook Block II Spacecraft, Volume 1, Spacecraft Description*, Image courtesy of NASA.)

also provided cooling water that circulated through the electronic equipment in the command module, preventing it from overheating.

On Earth, the equipment could be expected to remain quite cool by convection with the surrounding atmosphere. Energy conducted into a gas causes its molecules to speed up. This in turns lowers the density of the gas as the molecules spend more time further away from each other. The portion of the atmosphere nearest the equipment is obviously heated first, so it has a lower density than the portions next to it. In a gravitational field, this causes the denser air to sink and buoys up the less dense gas. This is how a *convection current* works. However, in a spacecraft on the way to the Moon, gravity plays a minor part and so the convection currents are very sluggish.[11] Hence the craft required a circulating coolant to conduct excess energy away. Excess heat extracted from the command module was radiated into space by two large radiators mounted on the service module.

The equipment that achieved all this environment control was a masterpiece of design. Redundant parts were used extensively to ensure reliability, yet the whole system occupied slightly more space than a large microwave oven.

Spanning the centre of the crew compartment were the three couches (Figure 5.8) designed to support the astronauts during the stresses of launch and splashdown and were connected to the command module structure by deformable struts. Looking in from the entry hatch on the sloping side of the capsule, the leftmost couch was occupied by the mission commander. The centre couch was for the command module

Figure 5.8 An artist's impression of the Apollo 12 command module interior. The command module pilot (CMP) is at his station in front of the large control panel across the capsule. In flight the centre couch would be stowed. During launch, the astronauts would recline on their couches and reach up to the controls. (Image courtesy of NASA.)

pilot who was responsible for navigating and manoeuvring the spacecraft during mission-critical moments (such as docking with and extracting the LM from the Saturn V third stage). The lunar module pilot took the right-hand couch. His main duty was to manage the sub-systems of the spacecraft. The seat portion of the centre couch could be folded to allow more room to move about. Sleeping bags were slung under each of the outer two couches.

The main instrument and control panel extended across the command module in front of the couches. This was slightly over 2 m long and less than 1 m deep with 90 × 60 cm 'wings' at both ends. The panel was divided into sections with specific controls and dials installed in front of the couches occupied by the responsible crew member (Figure 5.9).

The flight controls were positioned before the mission commander (CMD). His instruments informed him of the craft's altitude, velocity and attitude and he had access to the controls for propulsion, crew safety, splashdown and emergency detection of the capsule (in case the splashdown missed the target). One of the two navigation and guidance computer panels was placed in this section as well. Controls for manoeuvring the spacecraft (including a small joystick) were mounted on the arms of the mission commander's and command module pilot's couches.

The command module pilot (CMP) faced the centre of the panel. He could reach some of the controls mounted on his colleague's stations at his left and right as well as his specific instruments and controls directly in front of him. The latter included controls for managing the propellant for the command module manoeuvring engines, environmental controls, controls and instruments relating to the cryogenic storage of liquid oxygen and hydrogen in the service module[12] along with various general warning systems.

Finally, the lunar module pilot (LMP) had an area containing communications, controls and instruments related to the electrical power systems and fuel cells, data storage and propellant management for the service module engine.

Great care was taken in the design of the control panel and the instruments embedded in it. A lot of training was done inside command module mock-ups (and inside similar simulators for the lunar module) in which the crew rehearsed every aspect of their flight. Problems with the design of the panel were found during such simulations and as a result modifications were made based on the astronauts' recommendations. All the controls were designed to be operated by the crew wearing their spacesuit gloves. The most critical switches and controls were guarded with flip-up covers to prevent activation by accident. Some even had locks that had to be released before they could be used.

APOLLO COMMAND MODULE MAIN CONTROL PANEL

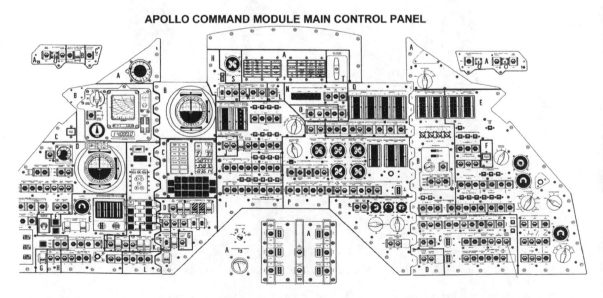

Figure 5.9 The central portion of the command module control panel. (From *Apollo Operations Handbook, Block II Spacecraft*, October 1969, Image courtesy of NASA.)

The efficiently planned volume of the crew compartment contained a number of equipment bays (essentially cupboards) to hold items needed by the crew as well as other equipment and electronics:

- The left-hand bay (to the left of the commander's couch) housed elements of the environmental control system with its displays and controls. It also included space for stowing the forward hatch while the connecting tunnel between CM and LM was open.
- The left-hand forward bay to the side and in front of the commander's couch contained more environmental control equipment as well as a water delivery system and clothing storage.
- The right-hand bay (to the right of the LMP couch) contained waste management controls and equipment (see Intermission 4), electrical power equipment, food storage and other electronic systems.
- The right-hand forward bay to the side and in front of the LMP couch was primarily a storage area with survival kits, medical supplies, optical equipment, the LM docking target and the bio-instrumentation harness equipment.
- The largest area was the lower equipment bay at the foot of the CMP couch. It served as a navigational station where the CMP could check and align the navigation systems with a telescope and sextant (see Intermission 3 and Figure 5.10). Most of the guidance and navigation electronics, the computer and a keyboard could also be found in this area along with the telecommunications electronics, five re-entry batteries and a battery charging system. The area was also used to stow food supplies and scientific and other equipment.
- The aft equipment bay was on top of the aft bulkhead separating the crew structure from the aft compartment. This area was used to store spacesuits and helmets, life vests, portable life support systems and other equipment. The docking probe and drogue assembly (see Section 5.2.6) were stowed here as well. The operations manual specifies that this was the area of the command module designated for the 'fetal cannister' (see Intermission 4).

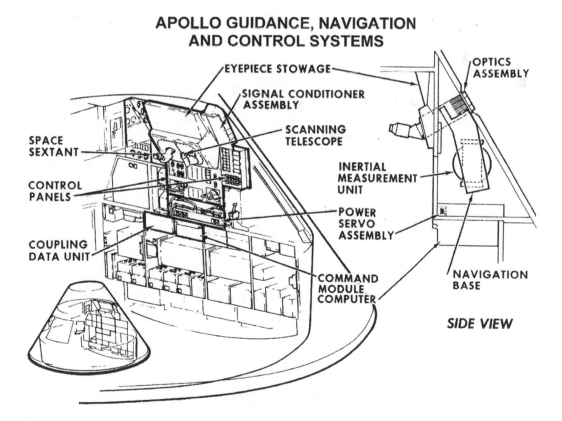

Figure 5.10 The location of the navigational optics within the lower equipment bay. (From *Apollo Training, Guidance and Control Systems, Block II,* Image courtesy of NASA.)

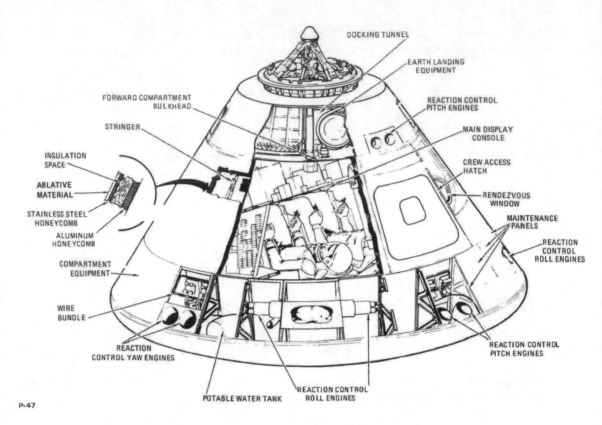

DOCKING TUNNEL

EARTH LANDING
EQUIPMENT

REACTION CONTROL
PITCH ENGINES

MAIN DISPLAY
CONSOLE

CREW ACCESS
HATCH

RENDEZVOUS
WINDOW

MAINTENANCE
PANELS

REACTION
CONTROL
ROLL ENGINES

FORWARD COMPARTMENT
BULKHEAD

STRINGER

INSULATION
SPACE

ABLATIVE
MATERIAL

STAINLESS STEEL
HONEYCOMB

ALUMINUM
HONEYCOMB

COMPARTMENT
EQUIPMENT

WIRE
BUNDLE

REACTION
CONTROL YAW ENGINES

POTABLE WATER TANK

REACTION CONTROL
ROLL ENGINES

REACTION CONTROL
PITCH ENGINES

P-47

Figure 5.11 An exterior view of the command module showing the reaction control thrusters. (From *CSM06 Command Module Overview*, Image courtesy of NASA.)

5.2.5 MANOEUVRING

While in space the combined command and service module units could be manoeuvred by employing one of two systems. Large course corrections, such as capture into lunar orbit, were achieved by firing the main *service propulsion system* (SPS). Smaller changes, docking manoeuvres and setting up the barbecue roll were done with the *reaction control system* (RCS) on the service module. However, once the service module was jettisoned (about 15 minutes before entry into Earth's atmosphere) the reaction control systems onboard the command module had to take over to fly the entry path through the upper atmosphere.

The command module was equipped with two independent systems of six thrusters (one served as a standby in case the primary system failed) distributed around the module in the aft and forward compartments (Figure 5.11). The thrusters used a hydrazine fuel and nitrogen tetroxide oxidiser combination (which is hypergolic) and each engine provided 400 N of thrust.

5.2.6 DOCKING

As the efficient docking of the CM and LM was key to mission success, it is hardly surprising that a great deal of careful design went into the equipment used to link the command and lunar modules (Figure 5.12).

The system consisted of a *probe* mounted on the end of the command module and a dish-shaped receptacle (*drogue*) on the lunar module. In essence, the two modules were manoeuvred together so that the probe slid down the inside slope of the drogue until the latches at the end of the probe engaged with the socket at the base of the drogue. When that happened, a partial dock had been achieved. The astronauts would then operate a probe retraction system, pulling the probe and the LM back towards the command module. At the contact between the two, 12 automatic latches in the CM docking ring engaged, forming an airtight seal between the craft (a hard dock).

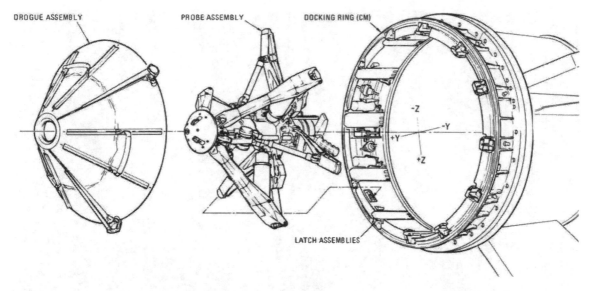

DROGUE ASSEMBLY   PROBE ASSEMBLY   DOCKING RING (CM)

LATCH ASSEMBLIES

Figure 5.12 The docking mechanism. (From *CSM 11 Docking Subsystem*, Image courtesy of NASA.)

Immediately behind the probe on the CM side and the drogue on the LM side were two halves of a tunnel that enabled astronauts to transfer between the two spacecraft. The halves were terminated by hatches on either side. Before the command module hatch could be opened, the pressure difference between the interior of the capsule and the tunnel had to be eliminated by operating a valve mechanism. After that was done, the CM hatch was removed and one of the astronauts checked the 12 latches, manually activating any that had not tripped. Then he connected an electrical cable between the CM and LM and floated down the tunnel until he reached the reverse side of the docking probe mechanism at the 'top' of the command module tunnel. The probe could be removed with the drogue still connected to the other side (remember that the 12 latches hold the spacecraft together!) and passed back to be stowed on one of the couches in the command module. Finally, he operated another valve to equalise the pressures between the tunnel and the LM before opening the last hatch, which was not removable: it hinged inward into the LM.

When it came time to separate the two vehicles for the landing, the two astronauts inside the LM would be passed the drogue by the command module pilot who would replace the probe and disconnect the cables. Then, he manually cocked all twelve latches (not releasing them, which could have been embarrassing) and closed the hatch. The separation was achieved by electrically releasing the latches from inside the LM.

The sequence was slightly different when the CM and LM docked for the second time, in lunar orbit after the LM ascent stage lifted off from the Moon. After hard dock was achieved, the command module pilot would open the CM hatch and remove the probe at his end while his colleagues in the LM opened their hatch and removed the drogue from inside the LM. When the LM was finally jettisoned, the release was achieved by disconnecting the whole CM docking ring with small pyrotechnic charges pushing the LM away from the CM.

The first docking that took place during a mission was scheduled for 25 minutes after the Saturn V third stage lit for the Lunar Orbit Injection burn. Explosive charges separated the command and service module combination from the third stage and the LM covers were blown open (Figure 5.13). The command module pilot used the thrusters on the service module to push the CSM ahead of the stage, rotate it and then come back to dock with the LM.

Optical aids ensured that the CMP had a means of accurately judging the speed of approach and alignment with the target on the LM. (During docking in lunar orbit, the CMP had to watch while the initial manoeuvring was done by the LM crew). Once a dock had been made, springs joining the LM's legs to the third stage were released and helped to push the combined spacecraft away from the stage. Thrusters were then used to pull the combination further ahead.

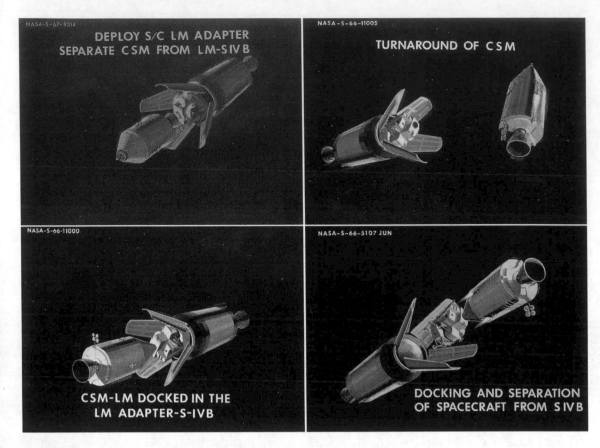

NASA-S-67-8314
DEPLOY S/C LM ADAPTER
SEPARATE CSM FROM LM-SIVB

NASA-S-66-11005
TURNAROUND OF CSM

NASA-S-66-11000
CSM-LM DOCKED IN THE
LM ADAPTER-S-IVB

NASA-S-66-5107 JUN
DOCKING AND SEPARATION
OF SPACECRAFT FROM SIVB

Figure 5.13 The sequence leading to LM extraction. (Image courtesy of NASA.)

The Apollo 14 mission was nearly brought to an end soon after it started when problems developed during docking. Like all CMPs, Stu Roosa on Apollo 14 trained hard for this manoeuvre and badly wanted the record for the smallest amount of fuel used. He manoeuvred the CSM into the LM drogue perfectly the first time, but it refused to make the initial latch. Three more attempts were made with no more success. Although in principle the CSM could coast along in front of the third stage indefinitely, the third stage was programmed to vent its remaining fuel into space after a certain time and it would be safer for the Apollo spacecraft to be well clear when that happened.

Onboard, the command module the mission commander, Alan Shepard, considered options such as depressurising the CM to retrieve the probe and attempt to fix it. As a final option, he was prepared to pull the spacecraft together by hand. The problem lay in the initial latch of the probe into the drogue. Shepard was sure the 12 main latches would trigger if he could just get the craft to join. In the end, mission control suggested that the astronauts make one more attempt by 'charging' at the LM, generating a harder impact that would contract the probe mechanism, bringing the latches into play. We will never know what the determined Alan Shepard might have done. The final 'charge' attempt secured a hard dock and the mission proceeded.

5.2.7 RE-ENTRY AND SPLASHDOWN

Command modules returning from the Moon entered the Earth's atmosphere at about 11 km/s and as a result led a trail of super-hot gas 200 km long. The drag on the capsule passing through the atmosphere at such enormous speeds caused it to decelerate at up to 6*g*. At the same time, the CM's aerodynamic qualities produced lift, in a similar manner to an aeroplane wing generating a net upward force.[13] The balance between drag and lift varied depending on the angle of flight.

If the capsule entered the atmosphere at too small an angle, the lift would climb quickly with little drag to slow it down, essentially because the spacecraft did not enter the dense atmosphere quickly enough. This would cause it to skip off the atmosphere into space, rather in the manner of a stone skipping off water. This would be an unretrievable situation.

If the capsule entered at too great an angle, the drag force would increase rapidly, slowing the spacecraft down. As lift is dependent on speed, the capsule would dip into the dense atmosphere too quickly. The drag would increase and there would be a danger of heat shield failure which would have killed the crew. As Figure 5.14 shows, the safe entry corridor was only 2 degrees wide.

The re-entry was flown automatically by the onboard computer, using the CM's thrusters to steer. The path was designed to descend into the dense air, then rise up briefly to give the crew and spacecraft a short respite from the extreme temperature and g force before diving again to complete the deceleration. If a failure in the guidance system occurred, the astronauts were trained to fly re-entry. Of course, no human could guide a craft as accurately as a computer, so the chances were that the crew would splashdown some distance from the target point, but at least they would be alive.

Having emerged from the enormous temperatures of re-entry, the capsule would have slowed to a more sedate 480 km/hour. The next stage of deceleration was achieved by parachute. First, the forward heat shield was ejected from the top of the command module (at about 7 km altitude) to allow the drogue parachutes to be deployed 1.5 s later. These 'chutes were 5 m in diameter and slowed the CM to 280 km/hour; they also ensured that the capsule was oriented at the correct angle for the remaining descent. When the capsule descended to 3 km altitude, the drogues were released and the pilot 'chutes took over, pulling the main three-parachute system from the CM. The main 'chutes were enormous (25 m in diameter) and the capsule hung 36 m beneath them. By the time the capsule reached the water, the main 'chutes had slowed the craft to 34 km/hour.

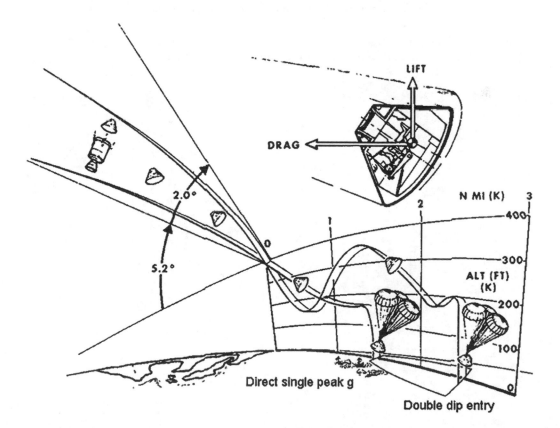

Figure 5.14 The entry corridor and sequence of events after aerodynamic deceleration. (Image courtesy of NASA.)

5.3 THE SERVICE MODULE (SM)

Length: 7.6 m	
Basic diameter: 3.9 m	
Overall mass: 24,523 kg	
Structure mass: 1,910 kg	

As its name implies, the service module provided the domestic services required by the crew during the journey to the Moon and back. Electrical power, oxygen, drinking water and other crew needs were produced and/or stored in the SM. As a result, the CM and SM were connected until 15 minutes before the CM hit the outer layers of the atmosphere at re-entry. The SM also provided the main propulsion system for the Apollo spacecraft via the large service propulsion system (SPS) engine needed for mid-course corrections, capturing Apollo in lunar orbit and breaking free of orbit again. Sixteen small reaction control engines (thrusters) on the SM could be used to manoeuvre the spacecraft during docking. The thrusters were grouped into four sets of four (quads) placed at 90 degree intervals around the circumference of the module. Each quad contained two thrusters that fired forward or backward along the module's axis and a further pair that fired clockwise or anticlockwise around the circumference. The system used the same propellant as the thrusters on the CM but were slightly larger as they produced 450 N of thrust. The SM thrusters were used to establish the barbecue roll. On later missions, the SM was equipped with a large instrument pack.

5.3.1 DESIGN AND CONSTRUCTION

The basic structure of the service module was a cylinder divided into six wedges surrounding the main engine (Figure 5.15, right). It was constructed from an aluminium alloy formed into a honeycomb bonded between inner and outer sheets.

A central cylinder 1.12 m in diameter (centre section) contained the engine and two nitrogen tanks. Running radially from this central core were six aluminium beams dividing the module into sectors and

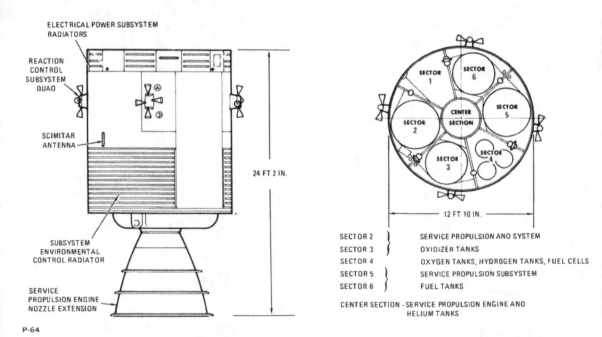

P-64

Figure 5.15 The interior of the service module. (From *Service Module Overview*, Image courtesy of NASA.)

The Apollo command and service modules

to which the external panels were bolted. At either end of the module were bulkheads. In addition, a heat shield across the bottom protected the module from the heat produced by the SPS.

The wedge sectors were not equal in size (Figure 5.15, right) and contained equipment and systems as follows:

- Sector 1 (50 degrees) was empty in the early missions. From Apollo 15 onwards, it contained instruments and a camera that the CMP used in lunar orbit to make mapping photographs, take radar soundings of the surface and perform other tasks.
- Sector 2 (70 degrees) contained part of a cooling radiator which overlapped into sector 3, a set of manoeuvring thrusters, fuel tanks for all the thrusters and the oxidiser sump tank for the SPS.
- Sector 3 (60 degrees) also had a thruster set on its outside panel and contained the SPS oxidiser tank. One of the two large cooling radiators for the environmental control system (continuing from sector 2) dominated the lower part of the exterior of this sector.
- Sector 4 (50 degrees) played a crucial role. It contained the three fuel cells that provided electrical power for the spacecraft. Driving these cells were the contents of two liquid oxygen tanks and two liquid hydrogen tanks. The oxygen tanks also delivered atmosphere for the CM.
- Sector 5 (70 degrees) contained the fuel sump tank. A thruster set was mounted on the exterior, and a second environmental control radiator started in this sector and continued into the next one.
- Sector 6 (60 degrees) contained the SPS fuel storage tank, the remainder of the second environmental radiator and the final thruster set.

With its domed lower heat shield, the CM had to sit inside a shallow dish at the top of the SM. Circling the top of the service module above the bulkhead was a ring (fairing) that formed the join between the command and service modules. Mounted round the fairing were the eight radiators used to vent the excess heat produced by the fuel cells into space (the electrical power subsystem radiators, Figure 5.15). Three tubes (one for each fuel cell) ran through each radiator.

In the vacuum of space, energy cannot be conducted or convected away. The only mechanism that works is radiation. The two larger (2.7 m^2) radiators around the lower part of the service module carried a water and glycol mixture from the CM to absorb heat from the cabin and electrical systems. Passing this liquid through pipes in the radiator allowed it to radiate energy into the cold of space. The coolant was then circulated back into the CM.

5.3.2 THE SPS

The service propulsion engine was just over 1 m long and its engine bell extended 2.7 m from the base of the service module and opened to a maximum diameter of 2 m. This crucial engine had to be ultra-reliable, with the fewest number of parts possible. If it did not 'light' at the appropriate moment, the spacecraft could not enter lunar orbit, or worse, break free of orbit to return to Earth. To ensure reliability, pumps were not used to transfer fuel and oxidiser to the thrust chamber. Instead, pressurised nitrogen from the two tanks mounted in the central core was used to force fuel and oxidiser from their storage tanks into their respective sump tanks and from there to the thrust chamber, a process that simply required valves to open. A hypergolic propellant mixture (fuel: 50% hydrazine and 50% unsymmetrical dimethylhydrazine (UDMH) oxidiser: nitrogen tetroxide) was used and 75% of the fully deployed module's mass was propellant.

The SPS was a potent engine, delivering 91,000 N of thrust. It could fire for a maximum of 8.5 minutes and be restarted 36 times. The propellant mixture had a specific impulse of 314 s and the SPS could provide a ΔV of 2.8 km/s.

5.3.3 FUEL CELLS (ELECTRICAL POWER SYSTEM OR EPS)

Electrical power for the command module was provided by three fuel cells (manufactured by United Aircraft's Pratt & Whitney Aircraft Division), each of which consisted of 31 cells connected in series, providing a total of 1.5 kW at 28 V. The cells operated in parallel and although only one was required to ensure the safe return of the spacecraft, mission rules vetoed a lunar landing if two of them failed. A command module that runs at less than 2 kW (equivalent to the power levels of many modern microwaves) represents impressive engineering.

The sub-cells contained a hydrogen compartment with an electrode made from nickel and an oxygen compartment with an electrode of nickel and nickel oxide. The electrolyte solution, a mixture of potassium hydroxide and water, provided a path to allow electrical conduction between the two compartments. At the hydrogen end, hydrogen gas from the cryogenic storage tanks reacted with hydroxide ions in the solution, a process catalysed by the electrode.[14] In chemical terms the reaction was:

$$H_2 + 2OH^- \rightarrow 2H_2O + 2 \text{ electrons}$$

the OH^- being hydroxyl ions from the potassium hydroxide solution. The electrons were drawn up the electrode to drive the current produced by the cell. Meanwhile, at the oxygen end, that gas (catalysed by the electrode) also reacted with the electrolyte solution. In this reaction electrons were drawn down the electrode, forming the return conduction path for the current from the electrical systems:

$$O_2 + 2H_2O + 4 \text{ electrons} \rightarrow 4OH^-$$

In addition to providing electrical energy, the two reactions produced heat and water as by-products. The warm water was piped to the CM where it could be used for washing and added to dehydrated food packages. Some of the heat was used to maintain the electrolyte solution at the correct temperature, and the rest radiated into space from the radiators around the top of the service module (Figure 5.15).

The tanks used to store the oxygen and hydrogen for the fuel cells had to be kept at cryogenic temperatures to maintain the gases in their liquid state.[15] These tanks had remarkable design parameters:

- very low energy loss to the surroundings: an Apollo's cryogenic tank could keep ice frozen for more than 8 years if placed in a room heated to 20°C;
- very low leakage of gases from the tanks: the same leak rate in a typical car tyre would flatten the tyre in 32.4 million years.

5.3.4 COMMUNICATIONS

The Apollo missions relied on a complex set of communication processes. In flight, the CSM and LM stack relayed telemetry, two-way voice communications and TV channels to the ground. At the Moon there were additional requirements including LM telemetry to ground, LM to CSM radio, LM to CSM radar and LM landing radar. On the lunar surface, systems for communication between astronauts and also between the surface explorers, the CSM and Earth were required. Finally, the ground teams needed communications equipment that enabled them to track the spacecraft during flight and throughout its journey around the Moon. The telemetry streaming down from the spacecraft carried detailed data about the conditions of the various sub-systems, including the crew who wore biomedical sensor packs to monitor their health.

The challenge of receiving radio from a distance as far as the Moon required technical advances on Earth. During the Mercury and Gemini programmes, separate systems handled voice communications, telemetry and tracking. The two-way radio and telemetry downlink utilised UHF[16] and VHF[17] bands respectively[18] while radar tracking relied on sending pulses in the C-band[19] from Earth to be relayed back by a passive transponder[20] on the spacecraft.

For Apollo, the distance to the Moon meant that passive radar ranging was not possible. Additionally, the extra requirements for TV coupled with the complexities of designing systems for each individual function (as for the Mercury and Gemini systems) drove the development of a more integrated approach. This resulted in the Unified S-band[21] system for Apollo which handled voice, TV, tracking and telemetry. A single carrier frequency was used each way and the Doppler shift of the received signal allowed the ground to measure the radial velocity of the craft. The data was combined by using sub-carriers which then phase modulated the main carrier. The system also allowed simultaneous information flow from two craft, although it did not completely replace all other telecommunication systems. VHF was also used for communications between the astronauts on the Moon and the LM, LM to CSM and communication via the lunar rover.

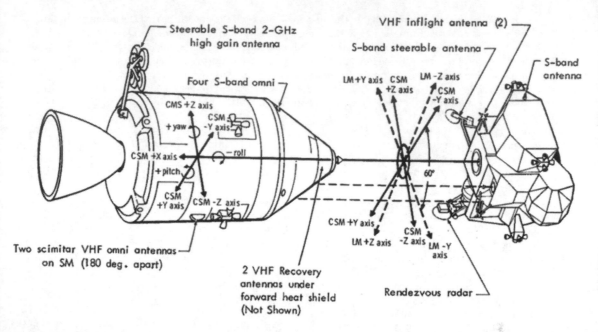

Figure 5.16 The antenna locations on the CSM and LM ascent stage. (From *Apollo 11 Press Kit*, Image courtesy of NASA.)

VHF between the LM and CSM used two 'scimitar' aerials, so named due to their resemblance to curved swords (Figure 7.5). One was installed on the LM and one on the SM. The main S-band antenna on the service module was a steerable collection of four 0.8 m diameter reflectors around a single square 0.28 m reflector (Figure 5.16). At launch, the antenna was folded down next to the SPS engine bell and protected by the LM shroud. After the CSM separated from the Saturn V third stage, it was deployed in its normal position at 90 degrees to the SM.

ENDNOTES

1. After the decision was made to use docking, gaining the necessary experience became a priority of the Gemini programme.
2. In other words, the ΔV needed to lift off the surface plus the ΔV for breaking out of lunar orbit and accelerating to Earth.
3. In some historical accounts, it has been suggested that Dolan was put on the track of LOR after visiting Langley where Houbolt and others were discussing this option.
4. Yuri Vasilievich Kondratyuk (1897–1942), notebooks compiled during military service on the Caucasian Front.
5. H E Ross presented the basic elements of LOR to a meeting of the British Interplanetary Society in London on November 13, 1948.
6. John C. Houbolt, Lunar Rendezvous, *International Science and Technology* 14 (February 1963).
7. In December 1966, NASA decided to conduct only one Block I spacecraft flight: Apollo 1. When the crew of that mission died in a fire during a test, their widows asked that Apollo 1 be reserved to designate the mission that never flew. As mission designs changed, so did the numbering system and the first manned Apollo command module flight was Apollo 7. The first flight of the Saturn V, with a basic Block I command module that incorporated many Block II features flew on November 9, 1967 as Apollo 4.
8. A honeycomb provides the most rigid structure with the lightest weight. You can make a simple honeycomb by taking several straws and glueing them together side by side. If the straws are then cut to 2 cm length and cards glued to the top and bottom (at right angles to the length of the straws), a very lightweight but rigid structure is the result.
9. The space shuttle was clad with specially designed ceramic tiles with very high thermal resistance to prevent conduction during re-entry. Modern crew capsule designs such as Space-X's Dragon also use ablative materials, albeit of more modern design.
10. The Pascal (Pa) is the SI unit of pressure; 1 Pa is the equivalent of a 1 N force acting over 1 m². Normal atmospheric pressure is about 100 kPa and cabin pressure is just over one-third of normal. The decision to use low pressure was partly based on engineering constraints. A much higher pressure would have made it difficult

to re-stock the cabin atmosphere in flight. The mission design cited the possibility that the astronauts would have to space-walk to cross from the LM to the command module should the two modules be unable to dock. Low cabin pressure was required to eliminate the need for an air lock (which would add weight and complexity).

11. During the Apollo 13 disaster, the astronauts had to turn off all the heating equipment in the command module as a means of conserving electrical power. The temperature dropped to that of a meat storage freezer. While attempting to sleep in the command module they discovered that their bodies could warm a thin layer of air surrounding them provided they did not move and disperse it. This would not have been possible if convection occurred.

12. On the ill-fated Apollo 13 mission, Jack Swigert, the command module pilot, was instructed to throw a switch that turned on a fan in one of the liquid oxygen tanks. This was a routine request from mission control as the fan was responsible for stirring the tank to prevent settling of the liquid. Unfortunately, a fault associated with the wiring to the motor caused an explosion that crippled the service module and very nearly cost the lives of the crew.

13. This happens because the air travelling over the top surface of the wing moves faster than air over the bottom surface. The faster the flow rate of a fluid, the less the pressure it exerts (see Chapter 3). Consequently, the greater pressure below the wing produces a net upward force. The same effect occurred as the command module entered the atmosphere tipped backward so its heat shield could protect the crew.

14. A catalyst is a substance that allows a reaction to occur but is not consumed in the reaction. Many chemical reactions will either not take place or proceed very slowly without the aid of a catalyst.

15. A given mass of gas can be stored much more effectively as a liquid as it takes up less volume.

16. UHF – ultra high frequency: 300 MHz to 3 GHz.

17. VHF – very high frequency: 30 to 300 MHz.

18. There was a high frequency (HF) 3 to 30 MHz backup radio system.

19. C-band: 4 to 8 GHz.

20. Passive transponders receive radar pulses and use their energy to re-transmit to the ground.

21. Unified S-band: 2 to 4 GHz.

The Apollo command and service modules

Intermission ③

Inertial guidance and computers

I3.1 THE NEED FOR A GUIDANCE SYSTEM

Imagine going on a journey and there are no signposts or maps to tell you where you are at any moment. The only landmarks are either so large they give you no accurate position or so small that you need a telescope with cross hairs to line up on them. How can you know where you are?

This is the problem faced by astronauts travelling to the Moon. They can see very big landmarks – the Earth, the Sun and the Moon – but they are so large it is difficult to use a telescope to provide a precise angular fix on your location. Other landmarks (e.g. stars) are small enough to yield accurate fixes but it is a very complex task to navigate to the Moon entirely on the basis of star sightings.

The answer to this navigation problem is to use the principle of *inertial guidance*, first developed by the Nazis as a means of directing missiles during World War II. The principle behind inertial guidance is very simple: you can always tell where you are, relative to an established starting point, by keeping a precise track of your speed and direction at all times. If you start from your house and walk at 3 m/s due south for 5 minutes, then turn 90 degrees left and walk at 5 m/s for 10 minutes, you should be able to determine precisely where you are after a few calculations.

A spacecraft has to do this in a slightly different way. It needs to 'know' how long its engines have been burning, the amount of thrust and the angle along which the thrust was directed. With this information a computer can calculate velocity changes, the direction in which the spacecraft is moving, and the distance it travelled along that direction. Consequently, it can keep track of where it is.

At first, the Apollo mission planners thought that all the guidance and control operations could be handled onboard by the ship's computer and the astronauts. However, the task proved to be so complex and required so much computer memory they ended up switching to ground-based guidance. The system onboard the spacecraft retained some independence, should contact be lost with the ground.

Trajectories were monitored from the ground by using a sophisticated system that measured shifts in the radio frequencies used to communicate with the ship. This Doppler effect is frequently heard when a fast-moving ambulance or police car passes the listener. As the vehicle approaches, the pitch of its siren increases and then decreases as it speeds past. The same effect applies to radio waves produced by a moving object. The equipment developed for Apollo was so sensitive it could even detect the slow rotation (barbecue roll) that helped regulate onboard temperatures. Computers on the ground were able to predict the precise Doppler shifts at any moment on a predefined trajectory and compare it to the frequency measured in order to check the spacecraft's path.

I3.2 GUIDANCE AND CONTROL SYSTEMS

The guidance and control systems for Apollo were designed to perform several functions:

- calculation of the craft's position and velocity (known as the *state vector* of the spacecraft);
- optical checks of the navigation platform;
- control and measurement of attitude (direction in which the craft points);
- control of the propulsion system;
- control of command module's path during re-entry.

Central to these functions was the *inertial measurement unit* – a gimbal-mounted[1] and gyroscope-stabilised block of machined beryllium. This formed a stable platform for an array of sensors that measured accelerations and rates of rotation about three perpendicular axes. The platform was locked down until shortly before liftoff, when it was released to move relative to the spacecraft. Given Newton's first law of motion, the stable platform tended to maintain a fixed orientation as the spacecraft rotated about it. This information, along with acceleration and rotation data was passed to the computer for use in tracking the craft's attitude and trajectory.

Periodically it was necessary to ensure that the stable platform had not drifted from its initial orientation. Even small errors in the data sent to the computer would create large deviations in trajectory over the distances required to reach the Moon. As the orbit was designed to fly them around the far side of the Moon within only 100 km of the surface, the crew were understandably keen to have an accurate fix at all times. To check the platform, the computer was set to rotate the spacecraft and angle the ship's telescope (or the sextant) to point to a specific star or alternatively a landmark on the Moon. If the target was not quite centred in the sight, the command module pilot navigating the craft would input the angular error into the computer, so the system could compensate.

On the first lunar orbital mission, Apollo 8, Jim Lovell was responsible for taking the sightings to align the inertial system. While in orbit about the Moon, he took fixes of several lunar landmarks to help define the ship's orbit and and serve as reference marks for the landing to come.

In this mode, the computer acted as an automatic aiming system. Lovell only had to feed the co-ordinates of the next crater or mountain into the computer and the sextant would be automatically positioned. The computer even tracked the ship's motion, so that he did not have to adjust the aim as they flew past landmarks. On his Apollo 13 mission, Lovell faced a serious problem when the explosion in the service module scattered debris around the craft. The glinting shards of scattered metal reflected sunlight and made it impossible to distinguish stars from chaff around the ship. Eventually the problem was solved by using larger points of reference such as the Sun.

Although the platform had a very sophisticated mounting, it did not have complete rotational freedom. If two of the gimbals supporting the platform happened to line up then the axes locked and the guidance platform was rendered useless. The cabin instruments included an 'eight-ball' or artificial horizon. Red circles painted on the device showed when the system was close to gimbal lock. Under normal circumstances, the spacecraft would not be manoeuvring so violently to create a problem. However, in the aftermath of the Apollo 13 explosion, the computer tried to compensate for the venting oxygen that pushed it off course by firing the thrusters on the service module. Some of the thrusters failed to work, causing inadvertently erratic motion that flirted with gimbal lock.

The same problem arose later, after the crew transferred to the lunar module. Lovell tried to use the thrusters to manoeuvre away from the cloud of debris that prevented him from getting a star fix. However, the lunar module's computer and the thrusters were calibrated for moving the LM without a very large and dead command and service module combination sitting on its back. The resulting motion nearly sent the lunar module's inertial platform into gimbal lock.

Routine inertial platform checks were generally performed prior to an engine burn. Another function of the computer was to control the burn activation, duration and direction of thrust. In some cases, this was critical; some burns had to be timed very accurately: a couple of seconds too long could mean crashing onto the Moon.

Directing an engine burn involved more than pointing a spacecraft in a specific direction and firing the engines. The service module's main engine was gimbal-mounted, so the direction of thrust could differ from the direction in which the craft pointed. Sensitive control was vital because the spacecraft's centre of gravity shifted as the propellant in the service module was consumed.

In one dramatic sequence from the movie *Apollo 13*, the crew had to fire the descent engine to correct their trajectory towards Earth. With the power off, the burn had to be timed and aligned manually. The timing was determined by one of the flight-certified Omega *Speedmaster* watches worn by Apollo astronauts. To get the burn direction right the crew aligned the craft with the demarcation line (the *terminator*) between day and night on the Earth's face. These events really happened. The only difference was that the ship shown in the movie fired its engine towards the Earth – a manoeuvre that would have increased the speed but failed to alter the angle. In fact, the ship was oriented vertically with respect to the direction in which it flew at the time, as the trajectory was too shallow.

The spacecraft's thrusters could also be controlled by the computer. These were used to rotate the ship about any one of three axes, thrust forward or backward (during docking or to 'trim' a major burn from the main engines), maintain a set attitude or set a rate of rotation (such as the barbecue roll). The lunar module, service module and command module had their own sets of thrusters, although those on the command module were not used until re-entry, after the service module was jettisoned.

13.3 THE APOLLO COMPUTER

The need for Apollo to carry an onboard computer was recognised early in mission planning although the reasons listed are quite illuminating:

- avoid the effects of hostile jamming that might cut off communication with the ground;
- gain experience needed for the interplanetary (e.g. Mars) missions to follow;
- prevent overloading the ground-based communications systems should many missions fly simultaneously.

In reality, the complexity of building and flying a spacecraft for travel to the Moon resulted in downgrading the onboard computer's role. All the major trajectory and burn calculations were performed on the ground and the solutions were transmitted to the crew for entry into their computer.

The contractor chosen for the Apollo computer system was the Charles Stark Draper Laboratory, where they had been working on guidance systems since 1939 when it was known as the Instrumentation Laboratory at the Massachusetts Institute of Technology (MIT). In 1970, the laboratory was renamed for its founder, and it became independent of MIT in 1973. Draper remains a highly successful company[2] and was awarded a contract under the Commercial Lunar Payload Services (CLPS) initiative supporting NASA's current plans to return to the Moon (see Section 9.1.5).

Once the mission mode was decided, it became clear that independent computer systems would be needed for the command module and the lunar module. The computers were identical in design but equipped with different software.

Normal in-flight manoeuvres could be backed up by ground-based computer control, but this did not apply to landing. The time delay between a signal departing the Earth and arriving at the spacecraft was enough to make the difference between a soft landing and a hard landing (crash). Hence, an onboard computer was required to manage the touchdown.

The lunar module was also equipped with a backup or abort computer. Its only job was to monitor the main LM computer and control an abort back to lunar orbit in case of a problem.

The early designs of the Apollo computers used transistor-based circuitry. However, the disadvantages of this technology led the team to become more innovative and daring. As early as 1962, the MIT group began to look at the newly developed integrated circuits (ICs) for use in computers. ICs were then only 3 years old and had not yet established a track record for reliability – always a major factor in NASA designs. Hence, MIT had to convince NASA that the advantages of using ICs were great enough to be worth the risk.

ICs of the 1960s were only pale shadows of today's microprocessors. They were single logic gates and MIT chose to construct the computer circuits from three-input NOR (Not Or) gates.[3] It would have been simpler to have a selection of logic gates but the use of a single gate improved overall simplicity and reliability. Almost 5,000 gates were installed in each Apollo computer. To put this in context, ~60% of the total IC production in the United States went into Apollo prototypes by the summer of 1963. Another example of Apollo's 'cutting-edge' computer design was the need for Raytheon (the contractor chosen to build the computer designed by MIT) to increase its production staff from 800 to 2,000 to cope with the demands of building 57 computers and 102 display keyboards.

The final computer was housed in the command module's lower equipment bay (see Chapter 5), and measured 61 cm by 32 cm by 15 cm with a mass of 32 kg. An equivalent computer in the lunar module was mounted in the cabin mid-section. All the circuits were placed in two trays of 24 modules. Each module housed two groups of 60 flat circuit boards with 72 pin connectors at one end. In turn, each circuit board held two ICs. Tray A held the logic circuits, power supply and interface electronics that connected the computer with other systems. Tray B contained the memory circuits, computer clock[4] running at 1 MHz (contrast that with modern smartphone chips that run at 2.49 GHz) and the circuits responsible for setting visible and audible fault alarms. Each unit was hermetically sealed.

The crew could input commands and read data from a set of identical display and keyboard units called DSKYS (Display–Keyboard, pronounced 'disky') (Figure I3.1). Two were installed in the command module: one on the instrument panel above the pilot's couch and the other at the navigation station in the lower equipment bay. One DSKY was mounted on the lunar module control panel.

The system for communicating with the computer seems crude by today's standards. For example, the early systems had no alphanumeric keys. On a modern computer, a request for a programme or operation is communicated by a mouse or keyboard command (or shouting!). On Apollo, a specific request was defined by a VERB, NOUN combination. Pressing the VERB button set the computer to read the next two digits typed as a base 8 code for a certain operation. NOUN instructed the computer to read the next two digits as the base 8 code for the component that the VERB operation would act on. ENTR performed somewhat like modern 'enter' or 'return' keys. The combination of a specific verb and noun activated a programme. If the programme required data entry by the crew, the VERB and NOUN lights flashed until all the expected digits were entered. The programmers provided the crew with a manual that listed all the combinations of VERB and NOUN, along with error codes and checklists.

As an example, VERB 3 7 ENTR instructed the computer to stop what it was doing and change to another programme. The next key presses (e.g. 3 1 ENTR) would switch to P31 (the rendezvous targeting programme). Various requests could then be made:

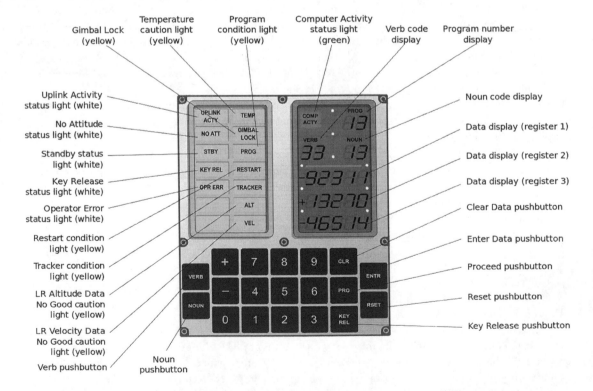

Figure I3.1 The standard display and keyboard unit in the Apollo computers. The black squares at the bottom are keys. The rectangles at the top left are alarm and other activity lights. At the right are computer displays that show the programme in use, memory locations (registers) and other data. (Image courtesy of NASA and Oona Räisänen.)

VERB 15 NOUN 18 – Display manoeuvre angles
VERB 06 NOUN 84 – Display velocity change for next manoeuvre

Memory requirements for the computer leapt as the data and programmes increased in complexity. The final configuration of the machine had 576k of RAM and 32k ROM.[5] Neither figure comes anywhere near the requirements for a modern operating systems (which in some ways says more about modern code than it does about a computer that could fly to the Moon!). The type of memory used was also very different from modern chips. The *core memory* of these early computers relied on electromagnetic principles.

I3.3.1 THE PRINCIPLE OF CORE MEMORY

A memory core consisted of many small 'donut' shapes made of *ferrite*, a ceramic–iron mix which is easily magnetised. The left side of Figure I3.2 shows a single memory unit. The two wires (X and Y) are responsible for magnetising the core, with half of the required current sent along each wire. The wires were linked to other cores, forming an array with rows and columns (Figure I3.2, right). The correct core could then be 'addressed' as it was the only one that had the current required to magnetise it (half from each wire). Each core could be magnetised in a clockwise or anticlockwise direction, depending on the sense of the current in the X and Y wires. A clockwise direction corresponded to '1' and anticlockwise direction meant '0'.

Reading the memory required sending a new current to an individual core via the X and Y wires. If this current was in the same sense as the one that magnetised the core, nothing happened. However, if the current in the read cycle was the reverse, it would re-magnetise the core in the opposite direction. The principle of electromagnetic induction states that a changing magnetic field will induce a current in a wire. As the core magnetisation flipped, it induced a pulse of current in the sense wire. If the flip was from clockwise to anticlockwise, the current pulse was in one direction. If the flip went from anticlockwise to clockwise, the current pulse was induced in the opposite direction. Sensitive electronic amplifiers monitored the sense wires, read the current pulse direction and converted into 1 or 0 as required. The key fact is that *changes* in magnetisation are read by sense wires; the process of reading memory automatically changes it.

The Apollo computer's RAM was constructed of memory core as described. However, the ROM followed a slightly different principle. Each ROM ferrite core recorded whole 16-bit computer words instead of single (1 or 0) bits. Each 16-bit word was represented by 16 wires. If the word was 1001000100010010, the first wire would miss the core, the second would go through it, the third and fourth would miss it, the fifth would pass through and so on. When the memory was read, only those wires that passed through the core carried a current pulse. A pulse read as 1 and the absence of a pulse as 0. This arrangement allowed each core to represent a word (in fact, each core on Apollo could record four words). The advantage of this *core rope* design was that the memory relied on how the wires were strung and so it could not be erased, even by a total power

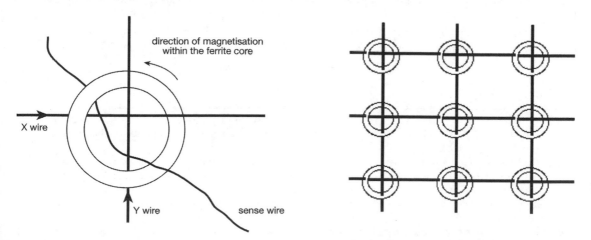

Figure I3.2 Left: a single memory core. A current in the X and Y wires produces magnetisation as shown in the core. Reversing both currents would reverse the sense of magnetisation. Right: Linking cores into an array. The sense wire would wind from one core to another.

failure. The design also allowed huge quantities of data to fit into a very small package. On the minus side, the memory blocks were difficult to build and it was impossible to correct any errors.

I3.4 SOFTWARE

If the computer design was remarkable for the time, then its software and engineers were no less so. Most of the software was written into rope memory, but some key programmes were changeable and so they had to be stored in core memory. The 'hardware people' developed the operating system for the guidance computer, along with its physical hardware. 'Software people' wrote the higher level code that governed the ship's operations.

Software engineering as a discipline did not exist in the early 1960s. The MIT team had to first develop the principles of code writing before they could get to the guidance and control software to fly to the Moon. Debugging and simulation were extensive to ensure reliability, such as the ability to recover from the programme alarms triggered during the Apollo 11 landing. In fact, no known software errors occurred during any of the manned Apollo missions. Despite the hard work and long hours needed, many of the engineers fondly recall the creative environment needed to meet the challenges of pioneering a new field.[6]

I3.4.1 SOFTWARE ENGINEERING 'ON THE HOOF'

As was routine, during the time between separation from the command and service module and the start of the powered descent to the Moon, the Apollo 14 LM crew carefully checked the systems and ran a 'dummy' landing through the computer. The first time they tried this, at the moment when the computer should have fired the descent engine it actually triggered the abort sequence. If it had been a real landing, the computer would have fired the craft into orbit again. Mystified, they ran the simulation again. This time the test proceeded normally for several minutes until the abort kicked in again. Worried that the landing would have to be called off, the crew had to wait while mission control tried to piece together a solution.

The most likely cause of the problem was a small ball of loose solder in the module's manual abort switch. The crew found that tapping the panel could clear the fault. The only way to work around the problem was to attempt to re-programme the computer to ignore the signal from the faulty abort switch. This was risky as it would have denied the crew an easy way to trigger the abort in case of trouble. Mission control had enough faith in the ability of the crew to proceed. However, the experts had less than 3 hours to rewrite the computer code.

Don Eyles, the MIT expert who programmed most of the landing software, was working late (at 1 am) in his office near the room in the MIT lab where the team monitored the mission. He rapidly came up with the basic idea which would solve the problem: set an indication that the abort was already in progress to stop the computer from triggering one. The team ran a couple of simulated landings in their system to iron out any bugs and then relayed the necessary 61 keystrokes to NASA, who then radioed the keystroke sequence to the crew for input into their computer.

By the time the new code arrived at mission control, Apollo 14's LM was passing out of radio contact behind the Moon. The signal was re-acquired only 30 minutes before the landing would have to be scrubbed. The crew typed in the new code with only 15 minutes left.

ENDNOTES

1. A gimbal is a type of universal joint that allows free rotation within a range of angles. Old-fashioned ship compasses were often gimbal-mounted so they could remain horizontal when a ship rolled in heavy seas.
2. https://www.draper.com
3. A logic gate is a circuit with a set of inputs and a single output. The output depends on the combination of signals presented to the input. The device operates like an electronically controlled gate that opens only when the correct code is entered. The simplest is the AND gate that turns on its output if all the inputs are ON as well. An OR gate turns ON when one or more of its inputs are ON. A NOR (not or) gate turns ON only when all its inputs are OFF.
4. This is not a time clock. Computer clocks provide the steady signal (rhythm) that times the sequences of operations to be carried out by the processing unit.
5. ROM is the acronym for Read Only Memory. ROM can be read but not erased or over-written. It is like a memory you can never forget. RAM denotes Random Access Memory that may be read, erased and over-written.
6. Excellent accounts can be found on the Draper website mentioned in note 2.

The lunar module

6.1 DESIGNING THE FIRST SPACECRAFT

Without meaning to suggest that the Block II command module was anything other than the technological marvel that it undoubtedly was, I have always felt that it would have been more interesting to work on the design of the lunar module (LM). After all, if we discount various space stations (on the grounds that they do not have a means of propulsion) the lunar module is still the only true manned *spacecraft* to have flown; in the sense of a vehicle intended to operate solely in the vacuum of space.

Free of the need to survive entry into an atmosphere, the design of the lunar module was dictated by expediency and the specific purposes of landing and providing a habitat on the Moon. The resulting design is certainly not aesthetically pleasing; many people feel it is ugly. However, the eyes of an engineer can see the purity of purpose and intelligent design (Figures 6.1 and 6.2).

Dimensions

Ascent stage height: 3.76 m
Descent stage height: 3.23 m
Diameter (diagonally across landing gear): 9.45 m
Earth launch weight: 14.5 tonnes
Pressurised cabin volume: 6.65 m³

6.2 THE ASCENT STAGE

A key issue identified in the development of the lunar module was visibility for the astronauts while landing on the Moon. An early design study by Grumman had a crew compartment reminiscent of a helicopter cabin, with four large windows forming a bubble-like enclosure (Figure 6.3). However, as the glass had to be thick to survive the pressure difference between the vacuum of space and the cabin environment, it added too much weight. Also, the heat lost into space through such a large area of glass would place severe demands on the cabin's environmental unit. Providing adequate visibility, especially downward, continued to be a problem until a re-think in another area provided an unexpected solution.

Designing seats that would support astronauts in spacesuits, provide adequate reach to the controls and be small enough to give sufficient room in the cabin for the astronauts to move around and clamber in and out of their spacesuits proved to be impossibly difficult. As a result, a radical step was taken: removing the seats – an unheard of departure in the context of a flying vehicle. However, given the zero-*g* during flight and the low gravity environment on the Moon, the engineers thought that the astronauts would be perfectly comfortable flying the lunar module standing up.

Resolving the seat problem solved the visibility problem as well. Two quite small triangular windows that angled back to the cabin were provided (Figure 6.4). As they were now standing, the crew were able to get closer to the windows and so had a much wider field of view, especially downwards. The size, shape and placing of these windows did much to fashion the unusual ascent stage 'face'. Furthermore, not having to worry about knee room for seated astronauts allowed the designers to shorten the cabin, saving weight.

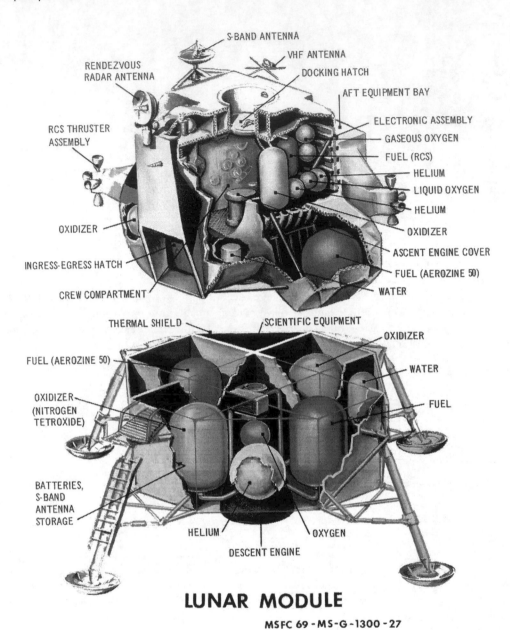

LUNAR MODULE

MSFC 69 - MS - G - 1300 - 27

FIGURE 6.1 The Grumman lunar module. This illustration shows the separate ascent and descent stages along with the locations of the various propellant and other tanks. (Image courtesy of NASA's Marshall Space Flight Center Collection.)

It was always thought that the lunar module would be, in essence, a skin wrapped around the tanks and equipment needed. Early designs used four propellant tanks: two for fuel and two for the oxidiser. This allowed the designers to place one fuel tank on either side of the engine and one oxidiser tank similarly arranged on either side, in a 'cross' formation. As a result, the tanks in each set balanced each other, keeping the stage's centre of gravity in line with the thrust direction of the engine. However, Grumman worried about the reliability of a design that required four tanks and associated plumbing, so a decision was made, with NASA approval, to carryout a $2 million re-design using only one tank for fuel and one for oxidiser. This shaved 45 kg from the mass of the stage but as the fuel weighed more than the oxidiser, the fuel tank had to be placed further from the centre of gravity than the oxidiser tank, giving the ascent stage its rather bulbous look on the right-hand side.

The lunar module

FIGURE 6.2 The Apollo 16 lunar module on the lunar surface. The image is cropped from a panorama. (Image courtesy of NASA and Apollo 16 crew.)

Figure 6.5 shows how the ascent stage's structure was subdivided. At the front was the cylindrical crew compartment, 2.35 m in diameter and 1.07 m deep. The metal decking provided about 1.5 m width of floor space and midway between the windows, the vertical clearance was 2 m. Facing forward, the mission commander stood on the left-hand side with the lunar module pilot on the right. The crew had armrests for use during powered descent. They could restrain in themselves in zero-*g* by a combination of Velcro strips on the floor to mate with those on the soles of their boots and tether cords that clipped to the spacesuits.

The commander and the lunar module pilot were both provided with a pistol-grip attitude controller for the right hand and a thrust control for the left. Mounted between these controllers on the LMP side was the Data Entry and Display Assembly (DEDA), used for configuring the abort computer.

In a similar position on the commander's side was a small control panel with 'Start' and 'Stop' buttons for the descent engine, a toggle switch that altered the rate of descent in 1 foot per second (30 cm/s) increments and controls for the lighting and the mission timer.

Between the commander and LMP stations and above the exit hatch sat another sub-control arrangement with the main guidance computer's display and data entry keypad. As this extended about 30 cm into the cabin, the astronauts had to be careful not to hit it with their backpacks as they left the lunar module.

The crew could access further instruments and controls on various other control panels vertically mounted on the forward wall of the cabin and on the bulkheads to either side. For example, the abort controls were on the commander's side of the panel between the windows. Similarly, the main 'contact light', which lit when any of the three probes extending from the decent stage legs touched the surface, was on the same panel at eye level. On the lunar module pilot's side of this panel were instruments and controls for

FIGURE 6.3 A 1963 design concept for the lunar module showing the helicopter-like crew compartment. The large glass windows were intended to provide adequate visibility for the astronauts during landing. The design assumed that the crew would be seated. (Image courtesy of NASA.)

the environmental system and the RCS thrusters along with various warning lights and a computer 'master alarm' indicator and cancelling button (a duplicate of the one on the commander's side). A small telescope for making star sightings hung from the ceiling over the computer panel with an eyepiece at eye level.

One interesting example of the astronauts' input into the configuration of the instrument panel involved the 'eight-balls'[1]. Such devices are generally used to display an artificial horizon as a reference for a pilot during flight. In essence, eight-balls are freely mounted spheres, which tend to remain level as the cabin rotates. However, as they are mounted into the panel, they do move along with the spacecraft! Comparing a line drawn across the ball's diameter with the reference scale on the glass cover of the instrument makes the angle of flight evident. The two eight-balls can be seen in their hexagonal mountings on the main panel in Figure 6.6.

Grumman initially installed them in the lunar module, assuming that the astronauts would wish to use them, but NASA had them removed. The astronauts demanded their return and they were re-installed. Then a NASA administrator asked why there was a development hold up on the LM and was told that there were problems such as the constant to-ing and fro-ing with the eight-balls. He had them removed again. The astronauts learned about the second removal and refused to fly the LM unless the eight-balls were in place. They were put back, with a kit for easy removal!

Behind the crew compartment was the midsection, marked by the docking tunnel at the top and two propellant tanks at the bottom. The internal part of the midsection was also pressurised and contained the guidance computer and environmental control system. Although the midsection was 2.3 m deep at eye level, space was curtailed by the overhead docking tunnel extending 40 cm into the cabin. The ascent engine cover below the docking tunnel reached knee height, leaving only 1 m of floor space in front of the ascent engine cover. The internal bulkheads were lined with equipment and storage compartments. The commander's side included stowage space for the backpacks (Portable Life Support Systems or PLSSs)

FIGURE 6.4 This beautiful photograph of the Apollo 17 lunar module's ascent stage clearly shows how thin the skin of the vehicle was in places. The triangular cabin windows and an outline of a suit helmet (right window) are visible. (Image courtesy of NASA.)

to be used on the lunar surface and the 'urine management system'. The opposite bulkhead housed the environmental control system.

Outside the cabin in the midsection region were the ascent engine fuel tank and its oxidiser tank. Also dotted about between the midsection's outer skin and the internal bulkheads were tanks for the liquid oxygen, liquid helium, gaseous oxygen and the propellant and oxidiser for the LM's thrusters.

The aft compartment of the ascent stage was used for equipment storage.

The entire structure was covered by a skin composed of two layers of Mylar for protection from small meteorites plus thermal insulation and a thin outer layer of aluminium.

Considerable debate surrounded the hatches used by the crew to enter and exit the spacecraft. The earliest designs assumed that the craft would need an upper hatch for docking with the command module when the LM was withdrawn from the Saturn third stage (a manoeuvre controlled from within the command module). However, when the ascent stage had to link again with the command module after return from the lunar surface, it was thought that the front hatch would be used. This time the lunar astronauts were responsible for the docking, and the need for a clear view during maneuvering necessitated the use of the front windows and hence the front hatch. This hatch also served as the means by which the astronauts would climb out of the LM onto the lunar surface.

However, as with many aspects of Apollo design, engineering a system to perform two jobs forced compromises that could not be tolerated. The forward hatch had to be reinforced to stand the stresses of docking and include a latching system similar to that designed for the upper hatch. These constraints made

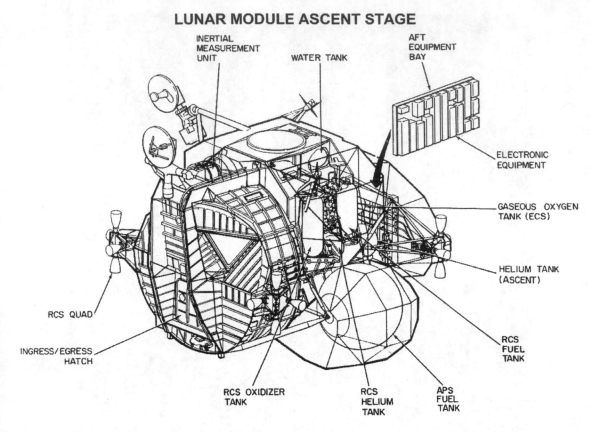

LUNAR MODULE ASCENT STAGE

INERTIAL MEASUREMENT UNIT

WATER TANK

AFT EQUIPMENT BAY

ELECTRONIC EQUIPMENT

GASEOUS OXYGEN TANK (ECS)

HELIUM TANK (ASCENT)

RCS FUEL TANK

RCS QUAD

INGRESS/EGRESS HATCH

RCS OXIDIZER TANK

RCS HELIUM TANK

APS FUEL TANK

FIGURE 6.5 A cut-away drawing of the lunar module ascent stage. The crew compartment is the cylindrical section at the front opening out into the midsection under the round docking hatch, which is visible at the top of the structure. (Image courtesy of NASA.)

it difficult for an astronaut wearing a spacesuit and a backpack to squeeze through the hatch. Eventually, a small window was added to the top of the cabin to allow the commander to see during docking to the upper hatch. This obviated the need for any complex mechanics in the front hatch. The window can be seen in Figure 6.5 as a distinct rectangular outline in the curved structure above the commander's station behind the left forward window.

After some trials with the fully suited astronauts, the shape of the forward hatch was changed from round (as needed for docking) to square, which was easier to exit through (square packs, round holes…).

Having the astronauts try things out was a vital part of developing the command and lunar modules. This was especially true when it came to solving problems relating to their descent from the LM hatch onto the lunar surface. During tests, Grumman arranged for a pulley system to support most of an astronaut's weight in order to simulate the Moon's gravity. Using this 'Peter Pan rig' the astronauts scrambled over mock-ups of the LM, to compare various ways of descending to the lunar surface. They even tested a knotted rope for climbing up and down. In the end, a flat plate (porch) was placed in front of the hatch and a ladder attached to one of the landing legs. Even the ladder design caused some debate. No one knew the depth of the fine dust on the Moon's surface, so the designers could not predict the extent to which the legs would sink. Consequently, the length of the ladder was determined by a somewhat educated guess!

6.2.1 THE ENVIRONMENTAL CONTROL SYSTEM

One of the most important sub-systems onboard the ascent stage was the Environmental Control System (ECS) which regulated the cabin's environment and supplied oxygen to the astronauts' suits while they were worn inside the LM. The unit was situated on the bulkhead behind the lunar module pilot's station, where it fitted into a space about 1 m high and 0.5 m wide.

The lunar module

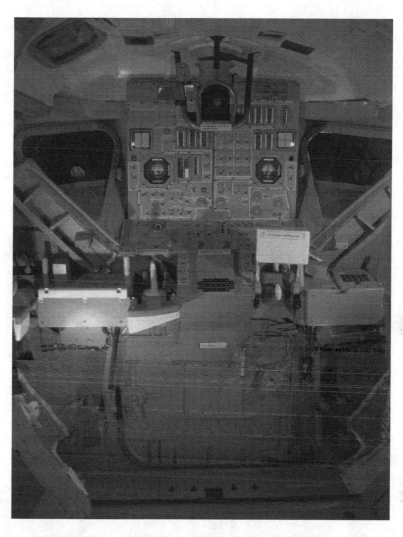

FIGURE 6.6 This image shows a mock-up of the lunar module interior. Some controls are labelled. An attitude controller can be seen on the right under the large label. The commander's thrust controller is on the left-hand side. The telescope is at the top centre. The hatch is visible under the control panel. The two hexagonal structures on either side of the centre line near the windows are the contentious eight-balls. (Image courtesy of Wikimedia Commons, Tyler Rubach.)

Normal Earth atmosphere at sea level consists of nitrogen (78% by volume), oxygen (21% by volume), water vapour and traces of other gases such as CO_2. The oxygen in this cocktail is by far the most important component for breathing. Working with pure oxygen enabled the cabin pressure to be reduced considerably, down to 33 kPa – roughly the same as the pressure exerted by oxygen in the atmosphere at sea level. Consequently, the LM's walls did not have to be extremely strong (saving weight) and the ECS had to handle only one gas, which reduced system complexity, saving more weight and improving reliability. Half-jokingly the astronauts were worried about the possibility of kicking a hole in the LM's skin!

The astronauts inhaled pure oxygen, with a little water vapour, from the cabin and exhaled CO_2 which, if not filtered out of the cabin, would build up to dangerous levels. As part of the ECS, lithium hydroxide canisters were used to filter out the CO_2. Fresh oxygen was supplied to maintain the cabin pressure.

The ECS was designed to maintain the cabin pressure at 24 kPa for at least 2 minutes should a 13 cm hole be punched in the cabin wall. Fortunately, this feature was never required. The only times that a moderate risk of puncture was foreseeable were during landing and ascent from the surface (debris could be kicked up from the ground in both cases) or during docking. During those manoeuvres, the crew wore their suits but did not inflate them as this would have restricted their freedom of movement. Two minutes was enough time for them to inflate the suits to survive depressurisation. The concern was not loss of oxygen so much as the loss of pressure to act against the body's internal fluids.

Depressurisation of the ascent stage prior to going outside (or throwing out the trash) could be done via one of two dump valves. One of them could be externally operated in case the ascent stage accidentally re-pressurised while the crew was on the surface, which would have made the hatch impossible to push open from the outside. Pulling on this valve could reduce the cabin pressure from 34 to 0.5 kPa in about 180 s.

6.3 THE DESCENT STAGE

The descent stage served three purposes: (1) it acted as a mount for the descent engine, which, as noted in Chapter 3, was possibly the greatest technological breakthrough achieved by Apollo; (2) it formed a stable platform from which the ascent stage could be launched back into orbit; and (3) it served as a storage bin for equipment, experiments, geological tools, replacement batteries, spare lithium hydroxide canisters, a 2 day food supply and in later missions, the lunar rover. Stowing equipment in the descent stage was advantageous. It left more room for the astronauts in the ascent stage and reduced the mass to be carried back into orbit; hence less fuel was needed for the ascent engine.

The descent stage was cross-shaped (Figure 6.7) and subdivided into square sections. The engine occupied the centre section with the propellant tanks arranged symmetrically around the outside to

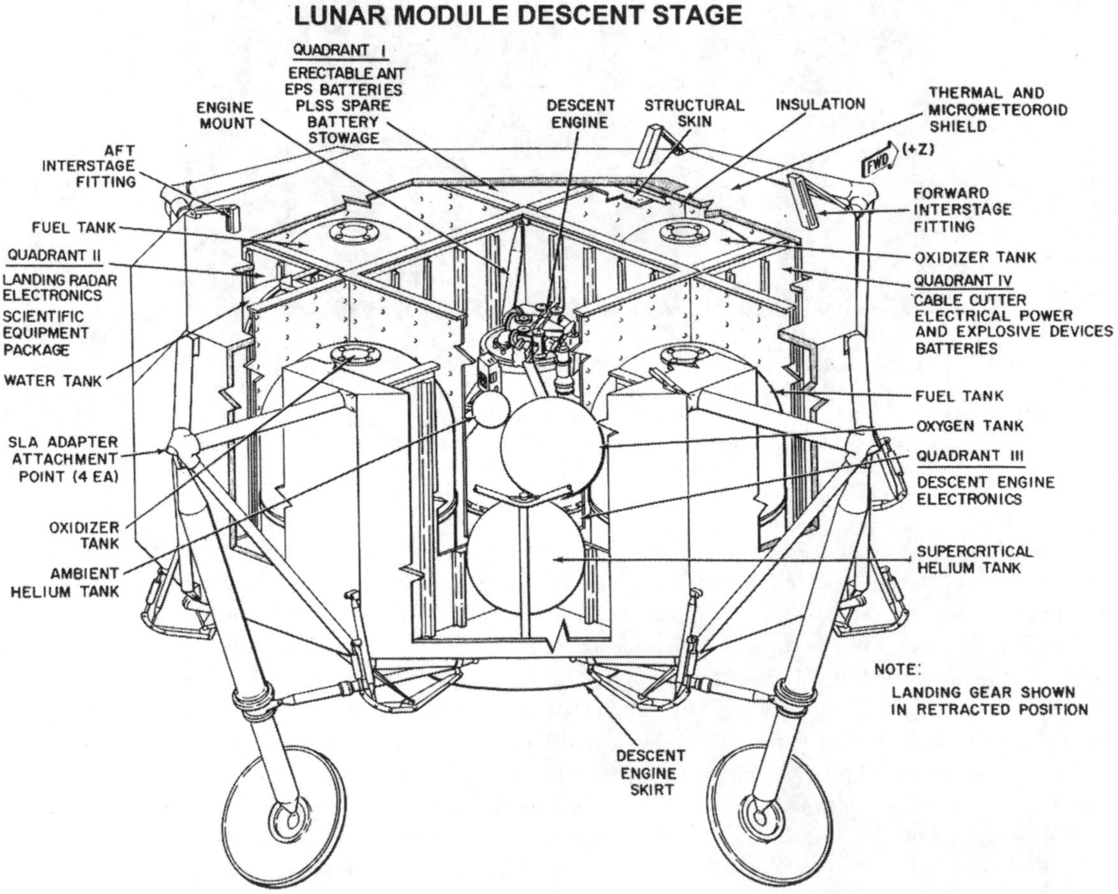

LUNAR MODULE DESCENT STAGE

FIGURE 6.7 The internal design of the descent stage showing the location of the descent engine. (Image courtesy of NASA.)

balance the weight distribution. The flat sides of the outer squares were used as attachment points for the landing legs. The remaining triangular sections, formed by joining the edges of the squares, were used as equipment stowage bays.

Space Technology Laboratories (STL) designed the descent engine to use a hypergolic propellant mix of Aerozine-50 (fuel)[2] and nitrogen tetroxide (oxidiser). The fuel flow rates were controllable manually or via the onboard computer system. As with the service propulsion system, the propellants were force-fed to the engine by pressurised helium from a storage tank. Control over thrust levels was achieved by regulating the propellant flow rates using adjustable valves in tandem with variable area injection ports to maintain uniform propellant velocity. The engine could also be gimballed up to 6 degrees.

6.3.1 POWERED DESCENT[3]

After the lunar module separated from the CSM, it flew ahead so that the command module pilot could give it a visual inspection (for example, he confirmed that the landing legs had extended correctly to clear the engine bell of the descent engine). The command module then fired its thrusters to perform a radial burn, placing it on an orbit that tracked 4 km above the lunar module's path during the descent. The command module came into line with the lunar module half an orbit later, on the far side of the Moon.

With the craft aligned, the descent engine fired to decelerate the LM and move it into an elliptical orbit, with the firing point becoming the apocynthion and the pericynthion having an altitude of 15 km at a point 14 degrees up-range of the landing site (Figure 6.8). This burn had to be performed on the far side of the Moon as the landing site was on the near side (for obvious communication reasons) and half an orbit is required to move from apocynthion to pericynthion.

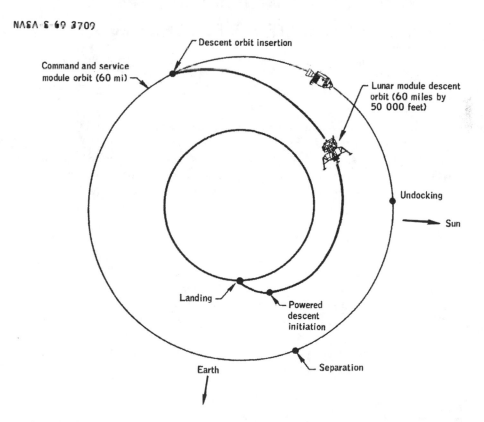

FIGURE 6.8 The phases of the descent process. First, the LM undocked from the CSM. The CSM then fired its thrusters to separate the craft. Descent orbit insertion was carried out on the far side of the Moon so that the lowest point of the descent orbit (half an orbit later) was on the near side. Then, the descent engine was fired again to enter the powered descent. (Image courtesy of NASA.)

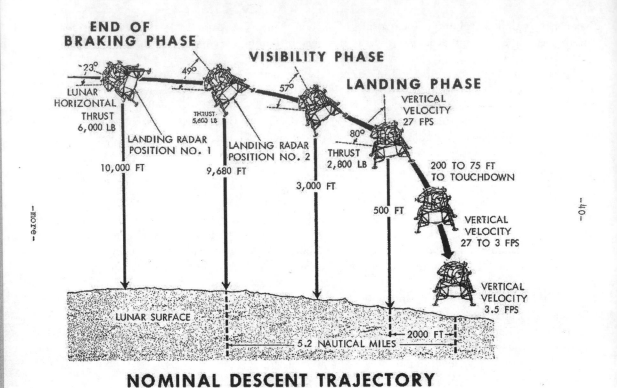

NOMINAL DESCENT TRAJECTORY FROM HIGH GATE TO TOUCHDOWN

FIGURE 6.9 The final phase of the powered descent. Although this graphic suggests that the lunar module was going to land sideways, the craft was flying on its back for the braking phase and pitched forward to allow the crew to see the surface during the final stage. (Image courtesy of NASA.)

Nearly 3 hours after the separation, the lunar module reached pericynthion and the descent engine fired again to break the craft out of orbit onto the powered descent trajectory. From this point to just before the landing, the computer flew the spacecraft. At the start of the burn, the LM was 482 km from the landing point and travelled at 1.7 km/s. The first phase of the burn was designed to slow the forward velocity essentially to zero. At the end of the braking phase, the computer fired the thrusters to tip the lunar module forward so that the crew could see the surface (Figure 6.9).

After making a visual assessment of the surface features, looking for landmarks and a safe landing spot, the crew could take manual control of the landing. With the correct throttle setting, the descent engine was able to counter the gravitational pull of the Moon, allowing the LM to hover. The commander could then use the thrusters to skim the craft around in order to find a landing spot. Finally, the craft entered a nearly vertical descent, slowing its vertical velocity from 8.2 m/s to 1 m/s prior to landing. Three of the descent stage's footpads were equipped with 1.7 m long probe wires. When they made contact with the lunar surface, the 'contact light' on the control panel lit as a signal to power down the descent engine. The LM could then fall to the surface.[4]

During the landing, the mission commander was responsible for flying the spacecraft while the lunar module pilot cross-checked the commander's actions and relayed instrument readings (such as altitude, rate of fall and horizontal velocity).

6.3.2 THE FLYING BEDSTEAD

Training for the mission-critical process of powered descent involved mission simulations inside full mock-ups of the LM and flying a most ungainly craft dubbed the 'flying bedstead' (Figure 6.10). Despite its rather odd appearance, the Lunar Landing Research Vehicle (LLRV) contained a sophisticated

FIGURE 6.10 The LLRV used in powered descent research. The pilot platform was designed to have a restricted view, similar to the experience in the lunar module. (Image courtesy of NASA.)

combination of sensors and computational hardware for the time. In lunar simulation mode, vertical acceleration was calibrated to the surface gravity on the Moon, to simulate landing under 1/6 *g*. The sensors could even automatically correct for wind gusts that deflected the vehicle – an adjustment that pilots on the Moon would not have to make.

Despite the level of automation, as an autopilot or stabilising system was not deployed, so some of the astronauts experienced rather 'close shaves' during training. On one occasion, Neil Armstrong was forced to eject at an altitude of 60 m, giving him only 4 s of parachute flight before impact with the ground. The accident investigation board found that the control thrusters had run out of fuel and that high winds during the test caused instability. This was to be the last flight of the LLRV before an improved Lunar Landing Training Vehicle (LLTV) was commissioned.

6.4 SPACESUITS

The lack of atmosphere on the Moon presents more problems than simple asphyxiation. The human body evolved in an environment that subjects it to an external (atmospheric) pressure of some 100 kPa. To compensate for this, internal body fluids exert an equal outward pressure on the skin. Consequently, an astronaut on the Moon deprived of suitable protection would explode due to the pressure difference between the inside of his body and the vacuum of space. Furthermore, the lack of atmosphere means that sunlight is not scattered or absorbed on its way to the lunar surface.

Over 50% of the energy reaching the Earth from the Sun is reflected or absorbed on its way through the atmosphere. On the Moon, this does not happen. Standing in bright sunlight on the lunar surface would be extremely hot (up to 127°C). In contrast, since the Moon has no atmosphere to convect and hence equalise the temperature, shadowed areas could be as cold as −157°C. On Earth, the atmosphere also provides protection from the Sun's ultraviolet light. Unprotected skin would burn rapidly on the Moon.

The role of a spacesuit is then to:

- produce and sustain a pressurised environment to support the body;
- provide breathing oxygen;
- prevent thermal losses and excess heating;
- protect against radiation;
- serve as a barrier against micrometeorites and scrapes against the lunar module or rocks on the surface that might otherwise puncture the suit;
- provide a means of communication between astronauts and, via a relay, to Earth.

Two versions of the spacesuit were designed for Apollo and manufactured by the International Latex Corporation. The command module pilot wore the 'intravehicular pressure garment assembly' while the lunar module pilot and commander had the 'extravehicular pressure garment assemblies'. The suits were identical, aside from the outer layers on the extravehicular type, which had extra thermal insulation and micrometeorite protection. The following descriptions relate to full Moon walking suits.

Each astronaut had three suits 'tailored' for them.[5] One was used in training (inevitably this one was subject to the most wear). Another served as the main suit for use on the mission. The third suit was provided as a backup in case a fault or rip developed in the main suit during the preparations for launch.

Working from the skin outward, the astronaut first wore an intricate set of knitted nylon Spandex 'long johns' through which a series of plastic pipes carried cooling water to extract excess body heat. During the normal in-flight routine, the crew wore Teflon cloth flight coveralls over the long johns.

A multiple layer pressure suit (Figure 6.11) covered the long johns. The innermost layer was made of lightweight nylon and designed for comfort. The second layer was a neoprene-coated nylon pressure

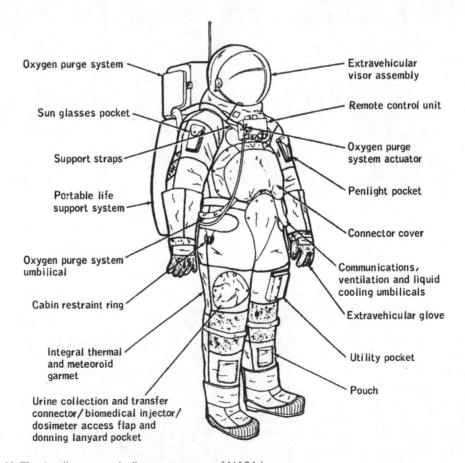

FIGURE 6.11 The Apollo spacesuit. (Image courtesy of NASA.)

bladder, skinned with a nylon layer to add strength. The bladder inflated when air was pumped into the suit at a pressure of 26 kPa, and along with the nylon layer, provided the chief resistance to pressure differences.

Working inside an inflated suit was rather like trying to move inside a rigid balloon. To achieve greater flexibility, rubber joints were provided at the knees, waist, elbows, shoulders and other areas. The joints were formed like bellows and reinforced with built-in restraint cables.

Next came a sequence of five aluminiumised Mylar layers (for thermal insulation) interspersed with four layers of Dacron. After this were two further layers of Kapton and beta marquisette (providing extra thermal insulation), a layer of Teflon-coated filament beta cloth (to protect against tearing and wear) and finally a white Teflon cloth layer designed to be flameproof.

The standard pressure suit came with shoes, but lunar overshoes were made for each of the astronauts who would be walking on the Moon. The exterior of the shoe was constructed from woven fabric and metal with a ribbed silicon rubber tread. The tongue was of Teflon-coated glass fibre cloth. The interior consisted of 25 alternating layers of the woven fabric and Kapton film, which provided insulation.

Gloves presented a problem for designers as they had to be flexible enough to enable the crew members to handle controls and tools. Two variations were developed. Black rubber gloves designed for launch were made from moulds of the astronauts' hands. The lunar surface gloves had silicon rubber fingertips to allow the wearer some degree of 'feel'. In both cases, the gloves were attached to the suit's arms by a combination of clips and locking rings.

The final component was the helmet. The basic launch helmet was a 'fishbowl' made from transparent polycarbonate with a small cushion behind the head. The astronauts on the Moon pulled hoods over their fishbowls and lowered gold-plated visors across the front to protect their eyes from the Sun (Figure 6.12).

FIGURE 6.12 Buzz Aldrin, lunar module pilot, on the surface of the Moon wearing an A7L spacesuit made by International Latex Corporation. The reflective coating on the helmet visor is evident. (Image courtesy of NASA and Neil Armstrong.)

During pre-flight checks, the crew waited in the suit-up room where small portable units provided oxygen for them in their suits. When launch control called them to the spacecraft, the astronauts carried the oxygen units to the van that took them to the pad. Once safely strapped into the command module, their suits were connected to the capsule's main environmental control unit.

Walking on the Moon required a great degree of freedom to move and use both hands. Consequently, while on the lunar surface, the astronauts wore sophisticated portable life support units on their backs.

6.4.1 BACKPACKS

Packing the variety of support systems the astronauts needed for walking on the Moon into a lightweight (353N on Earth) unit was a masterpiece of engineering (Figure 6.13). The backpack carried breathing oxygen (0.8 kg), fans to circulate it through the suit, the lithium hydroxide filters needed to extract CO_2, cooling water, a pump, a radio, a battery and the equipment to recharge the suit from the LM supplies.

The cooling water pipes in the astronaut's long johns were linked to a connector on the outside of the suit and from there via a hose to the backpack where pumps circulated water around the pipes without any significant loss. A second supply (called the *feedwater*) passed at a controllable rate through a heat exchange system where it was in thermal contact with water from the cooling circuit and conducted energy away, reducing the temperature of the water before returning it to the suit. The hot feedwater was allowed to evaporate into the vacuum. The backpack held about 5 kg of feedwater which was sufficient to provide cooling for about 8 hours of strenuous activity.

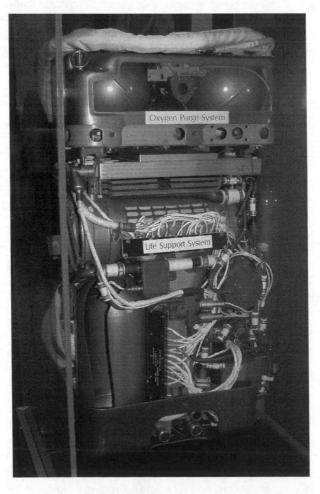

FIGURE 6.13 An Apollo-era backpack on display with the outer cover removed. (Image courtesy of NASA.)

The lunar module

For emergencies, each astronaut carried an 18 kg Oxygen Purge System (OPS) on top of his backpack. In the event of a backpack failure, suit puncture[6] or depletion of the main oxygen supply, this system could provide backup. The same system was intended to supply the suits should the LM be incapable of forming a hard dock with the CSM, in which case the astronauts would have to spacewalk to return to the command module. If the cooling system in the suit continued to work, the OPS could be set to provide oxygen at 1.8 kg per hour (about a 1 hour supply). If the suit's cooling system failed, the OPS had to be set to a high flow rate, yielding only about 30 minutes of breathing oxygen. On later missions equipped with lunar rovers (see Section 6.5), 1 hour's worth of driving time was built into the moonwalk plans for an emergency return to the LM. In that case, two suits could be linked together so that cooling water from one backpack could supply the other, allowing the slow-flow setting to be used.

After the last Moonwalk, and before leaving the Moon, the astronauts depressurised the lunar module, opened the hatch and threw out the trash, including the backpacks, to save weight in the ascent stage.

6.5 THE LUNAR ROVER

The Apollo lunar rover was commissioned by the Marshall Space Flight Center, then designed and built by engineers at Boeing. The vehicle was powered by two 36 V batteries although as usual with redundant design, one battery alone was sufficient to operate all the onboard systems. The rover had a mass of 210 kg and could manage a range of 65 km. It was capable of reaching a speed of 10 to 12 km per hour on level ground. In practice, the range was restricted to a 9.5 km radius to ensure a reasonable 'walk back' distance to the LM should the rover be immobilised.

If you take into account the tasks that had to be performed to ready the rover and shut it down after use, the astronauts did not gain much time by using it to travel over the surface rather than walking. On the other hand, by using the rover, the explorers consumed less oxygen, did not feel as tired and were able to carry more equipment and return with more samples.

The combination of low lunar gravity and a surface covered in fine dust provided interesting challenges for the designers of the rover's suspension. The wheels were made from a strong woven zinc-coated piano wire mesh riveted with titanium treads to form chevron-shaped tyre grooves (Figure 6.14). They

FIGURE 6.14 The Apollo 15 lunar rover in its final parking place on the lunar surface. (Image courtesy of NASA.)

were independently driven by electric motors (about 1/4 horsepower each) and separate steering systems controlled the front and back wheels.

While in motion, the rovers kicked up a great deal of dust and so needed 'mud guards' or fenders over the top of each wheel. I remember watching a TV transmission showing the astronauts driving about the lunar surface, testing a rover. The scattered dust fell back to the ground in curiously slow motion due to the low lunar gravity.

During the first Extravehicular Activity (EVA) of the Apollo 17 mission, one of the astronaut's geological hammers caught in a back tyre fender and broke part of it off. Lunar dust has not been weathered smooth. As a result it is far more abrasive than its terrestrial equivalent – and more of a hazard when it contaminates suits, experiments and equipment. Dust kicked up by a rover without a fender presented a serious issue. Furthermore, the dark dust would absorb sunlight effectively and raise the temperatures of instruments, potentially to the failure point. While the two men slept on the Moon, mission control worked up a solution which involved taping together four maps from the astronauts' kits and affixing them to the damaged fender using clamps from the optical alignment telescope on the LM (Figure 6.15).

Remarkably, for the time, the rover was equipped with an onboard navigation system. On earlier missions, astronauts discovered that finding their way about on the lunar surface was often harder than they expected. The combination of crystal-clear viewing (no atmosphere to blur distant objects) and the total uniformity of colour made judging distances very difficult. It was easy to spot the LM on the surface, as it was highly reflective, coloured differently from its surroundings and was 7 m tall, so the astronauts were never in any danger of getting lost. However, a great deal of planning went into the EVAs from a geological standpoint. Scientists on Earth wanted samples taken from precise locations. The rover's navigation system helped achieve that.

FIGURE 6.15 The Apollo 17 fender repair using maps, clamps and tape. (Image courtesy of NASA.)

FIGURE 6.16 The Apollo 17 lunar module ascent stage takes off, leaving the descent stage on the surface, as filmed from the lunar rover. (Image courtesy of NASA.)

Before a trip started, the rover's system was reset at a point near the LM. As the astronauts drove off, the system counted rotations of the rover's wheels to determine distance and used a gyroscope for directional information. In this way, it was able to pinpoint the rover's location relative to the starting point with an accuracy of ~100 m. However, to effectively help the crew navigate to a desired location, the starting point had to be accurate. On the missions that deployed the rover, the astronauts managed to unambiguously recognise some surface features such as craters that enabled those on the ground to fix a location from which they could direct the EVAs.

A remote-control TV camera, movie camera and two aerials were mounted on the rover's chassis. The TV camera was motorised and could be directed by the controllers on Earth. Along with providing stunning pictures of the astronauts' activities during EVAs, the camera also panned about the lunar panorama. The most memorable view relayed to Earth, however, was the launch of the lunar module ascent stage. The rover was parked a safe distance away and the camera tracked upward as the ascent stage made its way into orbit (Figure 6.16).

6.6 THE ASCENT TO ORBIT

With their stay on the lunar surface completed, the two astronauts could stow their equipment and rock samples and fire up the lunar module's ascent engine to return to orbit. The powered ascent was divided into two phases: vertical ascent and orbital insertion. The vertical ascent was a flight straight upward from the lunar surface, to ensure that the LM cleared the terrain before it pitched over so that the ascent engine could accelerate the horizontal motion up to orbital speed. Pitch-over started when the LM's vertical ascent rate reached 15 m/s, some 10 s after the burn started and at an altitude of 76 m. The ascent engine continued to burn ~7 minutes, achieving a ΔV of 1.8 km/s. Upon completion of the powered ascent, the LM settled into an orbit which was 16.7 km by 81.6 km relative to the surface. The ascent engine was not used again. All the remaining manoeuvres were carried out with the reaction control system's thrusters.

With the fine details of their orbit available, the crew would use the LM's main computer to calculate the final adjustments leading up to rendezvous with the CM in its circular 111 km orbit. When the LM reached apolune, the RCS thrusters fired to achieve a ΔV of ~15 m/s which would move the LM into an orbit about 28 km below that of the command module (84.3 km by 81.9 km). In this lower orbit, the LM would be travelling more quickly than the CSM and caught up at a rate around 0.072 degree per minute. Additional burns of the RCS thrusters took place when the LM and command module were closer together in order to slow the catch-up rate and then dock the two modules. The entire sequence from the start of the first RCS burn took about 3.5 hours.

ENDNOTES

1. The eight-ball is an important sphere in billiards (apparently…).
2. A 50:50 mix by weight of hydrazine and unsymmetrical dimethylhydrazine (UDMH).
3. The details of powered descent varied from flight to flight. This account is based on the landing of Apollo 11.
4. In practice, the engine was not always cut off at this point.
5. The modern EVA suits worn on the ISS are more universal. Only the gloves are individually tailored.
6. Of course, this would not repair the hole; oxygen would still leak out, but the system could maintain the pressure while the astronaut returned to the LM.

Intermission (4)

The three 'ings'

I4.1 EATING

> Balanced meals for five have been packed in man/day over-wraps, and items similar to those in the daily menus have been packed in a sort of snack pantry. The snack pantry permits the crew to locate easily a food item *in a smorgasbord mode* without having to 'rob' a regular meal somewhere down deep in a storage box.
>
> *Apollo 11 Press Kit, page 123[1] with author's emphasis*

This extraordinary quote makes an Apollo mission sound like a cross between a Sunday school picnic and military exercise. Food for the astronauts was planned carefully to provide a balanced intake over the duration of a flight (Table I4.1.1). The restrictions of zero-*g* dictated that food selections were packaged as freeze-dried and rehydratable, wet-packed or potted to be eaten with a spoon.

The command module had a dispenser for normal drinking water and two taps that produced measured amounts of water warmed to 68°C or 13°C by excess heat from the fuel cells. Water was injected into food packages which the astronauts 'kneaded' for several minutes to ensure a good mixture. Finally, the top of the package was cut off and the contents squeezed into the mouth. After the meal, the astronauts placed germicidal tablets into the empty packet to prevent fermentation of the remains, rolled up the bag and stowed them in a waste compartment.

I4.2 SLEEPING

Most astronauts reported sleeping problems because they missed the support of pillows and the weight of covers on their bodies. The sleeping crew tended to float about the cabin, which is why sleeping bags that could be tethered in place were provided.

The situation improved slightly on the Moon where gravity was of some help. However, astronauts in the lunar module faced their own problems when it came to sleeping arrangements. On the early missions, the crew had to spend all their time on the Moon in their spacesuits. The inflexibility of these bulky suits made it difficult to get comfortable. By Apollo 15, a re-design made the suits easier to remove in the lunar module's cabin and from then on the crew could strip to their long johns to sleep. On Apollo 11, Neil Armstrong and Buzz Aldrin slept perched on the ascent engine cover and curled up on the floor respectively. Later missions carried beta cloth hammocks, one of which could be strung across the cabin and the other attached above it

Table 14.1.1 Neil Armstrong's menu on the Apollo 11 flight

MEAL	DAY 1,5	DAY 2	DAY 3	DAY 4
A	Peaches Bacon squares Strawberry cubes Grape drink Orange drink	Fruit cocktail Sausage patties Cinnamon toasted bread cubes Cocoa Grapefruit drink	Peaches Bacon squares Apricot cereal cubes Grape drink Orange drink	Canadian bacon and applesauce Sugar-coated corn flakes Peanut cubes Orange–grapefruit drink
B	Beef and potatoes Butterscotch pudding Brownies Grape punch	Frankfurters Applesauce Chocolate pudding Orange-grapefruit drink	Cream of chicken soup Turkey and gravy Cheese cracker cubes Pineapple–grapefruit drink	Shrimp cocktail Ham and potatoes Fruit cocktail Date fruitcake Grapefruit drink
C	Salmon salad Chicken and rice Sugar cookie cubes Cocoa Pineapple–grapefruit drink	Spaghetti with meat sauce Pork and scalloped potatoes Pineapple fruitcake Grape punch	Tuna salad Chicken stew Butterscotch pudding Cocoa Grapefruit drink	Beef stew Coconut cubes Banana pudding Grape punch

Note: Armstrong's menu on Day 1 consisted of meals B and C only. Stored along with the food were toothbrushes and tubes of toothpaste. Supplies of wet wipes and tissues were also stowed about the spacecraft.

along the cabin's length. After the crew spent several days in the command module, the unfamiliar sounds of the lunar module systems and the excitement of being on the Moon made it hard to rest.

14.3 EXCRETING

Of all the natural functions that the crew had to carry out onboard Apollo, 'waste management' was the most primitive and unpleasant.

Urine was dealt with by a hose with a condom-like fitting at one end. The principle was to attach the hose, open a valve, then urinate into the vacuum of space where the liquid would freeze immediately into shimmering crystals. Apparently, the sight was quite spectacular. In practice, however, the unit was not that easy to operate. Aside from the psychological problems involved in connecting part of one's anatomy to the vacuum of space, opening the valve created a pressure difference that drew things into the hose. Shutting the valve tended to trap anatomical extrusions in the hose as well.

If this aspect of spaceflight was not the most convenient, then solid waste management was even worse. The mechanism consisted of a top hat shaped bag with a sticky rim. The idea was to attach the bag in the proper position and proceed, while ensuring that the material remained in the bag—not an easy job in the zero-g environment of a spacecraft. The final delightful task was to place a germicidal pill into the bag and to knead the contents into a good mix. The bags were then stored for analysis back on earth.[2]

The toilets devised for the space shuttle and the ISS were far more sophisticated and could be used by both sexes. They were similar to those found on modern jet airliners, with the addition of spring-loaded thigh restraints!

ENDNOTES

1. https://www.hq.nasa.gov/alsj/a11/A11_PressKit.pdf
2. Nice work if you can get it …

7

The missions

7.1 THE GREAT TRAGEDY: APOLLO 1

Space flight will never tolerate carelessness, incapacity and neglect. Somewhere, somehow we screwed up. It could have been a design in build or in test, but whatever it was, we should have caught it. We were too gung-ho about the schedule, and we locked out all of the problems we saw each day in our work…

From this day forward, Flight Control will be known by two words: Tough and Competent. Tough means we are forever accountable for what we do or what we fail to do. We will never again compromise our responsibilities… Competent means we will never take anything for granted… Mission Control will be perfect. When you leave this meeting today you will go to your office and the first thing you will do there is to write Tough and Competent on your blackboards. It will never be erased. Each day when you enter the room, these words will remind you of the price paid by Grissom, White, and Chaffee. These words are the price of admission to the ranks of Mission Control.

Gene Kranz's *speech to the mission controllers 2 days after the Apollo 1 fire*

On January 27, 1967 a tragedy struck the preparations for Apollo that was unprecedented in the history of the manned space flight programme. Three astronauts, Gus Grissom (Mission Commander, Mercury 2, Gemini 3), Ed White (Command Module Pilot, Gemini 4) and Roger Chaffee (Lunar Module Pilot) were in full spacesuits strapped into their seats in a Block I command module atop a Saturn Ib booster on Pad 34. They were there for a countdown dry run. It was to be a complete test of the systems both onboard the spacecraft and in the control rooms prior to their launch on the first manned Apollo mission scheduled for February. The tests had been long and arduous. The command module revealed several bugs in its components and a maddening fault with the communications systems. However, for one brief terrible moment, contact with the control rooms was clear and unambiguous. At 6:31 pm, about 5½ hours into the test, a short transmission from the command module froze the blood of all who heard it: 'Fire!'

Within moments, this first hint of trouble was followed by 'we've got a fire in the cockpit' and then, with mounting urgency 'We've got a bad fire…. We're burning up.' Television monitors viewing the outside of the command module hatch showed the crew trying to release the mechanism. On the pictures, a bright glow inside the capsule and flames licking out around the hatch could be seen through billowing smoke. Eventually, the hatch window was completely obscured by the smoke inside. Meanwhile, five people outside the spacecraft attempted to fight the intense smoke, heat and flames to reach the crew. They finally gained access to the inside of the command module *5 minutes* after the first indications of trouble. They were too late. All three astronauts died on the launch pad.

Postmortems revealed that the spacesuits protected the crew from most of the fire. They had second- and third-degree burns, but they were not fatal. Asphyxiation killed the astronauts. The suits were connected to the command module systems by breathing hoses that relayed a pure oxygen atmosphere to the crew. When the fire burned through the hoses, the smoke and other fumes entered the spacesuits. The astronauts died in moments.

In the weeks and months that followed, strenuous investigations into the incident revealed the likely cause of the fire but also made management aware of complaints that the astronauts had been making for some time. The NASA internal review board issued a summary report[1] on April 5, 1967 that stated:

Although the Board was not able to determine conclusively the specific initiator of the Apollo 204 fire, it has identified the conditions which led to the disaster:

- A sealed cabin pressurized with an oxygen atmosphere.
- Extensive distribution of combustible materials in the cabin.

- Vulnerable wiring carrying spacecraft power.
- Vulnerable plumbing carrying a combustible and corrosive coolant.
- Inadequate provisions for the crew to escape.
- Inadequate provisions for rescue or medical assistance.

Having identified the conditions that led to the disaster, the Board addressed itself to the question of how these conditions came to exist. Careful consideration of this question leads the Board to the conclusion that in its devotion to the many difficult problems of space travel, the Apollo team failed to give adequate attention to certain mundane but equally vital questions of crew safety. The Board's investigation revealed many deficiencies in design and engineering, manufacture and quality control.

A pure oxygen atmosphere was specified instead of an air-like mixture of two or more gases as that would have involved heavy and complex additions to the cabin environment systems. While on the pad, a pure oxygen environment at greater than atmospheric pressure had been chosen to flush any contaminants from the command module. Many objects burn readily in oxygen atmospheres. Almost anything will burn in a cabin soaked in pure oxygen at greater than atmospheric pressure for 5 hours.

The most likely source of the ignition was under Grissom's couch, where bundles of wires ran across the floor of the capsule. These wires had been moved, squeezed, bent, trodden on and generally abused during building and re-engineering the capsule. So many engineering changes occurred over the months it took to build Capsule 012 (as it was known to the contractors) that no one had a clear idea of the exact configuration of the spacecraft – a fact that amazed investigators. Indeed, the team who stripped down the charred Apollo 1 capsule found a spanner lodged in the wiring loom.

In the rush to get to the Moon by the end of the decade, short cuts were taken. The design of the Block I spacecraft started before the mission mode had been finally decided so Apollo was a work in progress. Changes were made 'on the hoof' and no records were kept to indicate what was done by whom.

It is likely that one of the electrical wires had damaged insulation. A spark occurred and the inside of the capsule turned into a blowtorch.

Incredibly, with the flames surrounding them, the crew reacted calmly and exactly as they had been trained. Grissom tried activating the depressurisation lever (already in flames) to vent the cabin atmosphere. White twisted in his seat to start opening the hatch. Chaffee attempted to contact the controllers and then turned to help White. Inside the command module, metal pipes melted and dripped onto the floor. Opening the hatch under such conditions was a physical impossibility.

The hatch consisted of an inner hatch that opened into the CM, an outer hatch that pivoted outward and finally a third hatch that was part of the boost cover.[2] Under normal circumstances, this combination could not be opened in less than 90 s and more likely required 2 minutes. The crew died within seconds of the start of the fire.[3]

In the aftermath of this disaster, confidence in the Apollo program was severely damaged. The nature and wisdom of the race to get a man to the Moon were questioned and the program faltered. 'Heads rolled' within NASA and on the contractor's management team. The start of the recovery was probably triggered by the publication of NASA's internal report. Criticism had been directed at NASA and Congress for not holding independent inquiries. In the end, people marvelled at how frank the NASA report was.

Just as a single source of ignition within the command module could not be isolated, no single person responsible for the disaster could be blamed. The problem was one of culture and pressure. For months, the astronauts had complained about design changes. It was not possible to keep their traning simulators up to date with the updates installed by the engineers building the real craft. Ironically, the backup crew for Apollo 1 were sealed inside a second Apollo command module doing their own test at the time of the fire. After a period of maddening frustration with the faults onboard their capsule, (at one point the hatch fell on an astronaut's foot), that test was called off.

Aside from not being able to keep up with the changes, the astronauts were unable to establish working relationships with the engineers. Many of the crew members in line to fly Apollo missions were experienced astronauts from the Gemini program. McDonnell Douglas built the Gemini capsule and a good rapport between the astronauts and the engineers made the job of development much easier. The contractors listened to the astronauts who would fly the spacecraft and made the changes requested.

There was no doubt that the people at North American Aviation were just as competent as those who built Gemini, but they had never built a spacecraft. The astronauts could not find a clear line of complaint to pursue in order to get results. Their criticisms were not confined to North American; they found the NASA designers just as unhelpful. Frustrations built up because the experienced astronauts felt that they were not being listened to. However, in the end, they recognised that the deadline had to be met and their 'right stuff' spirit (most of them had been test pilots) mandated that NASA and North American 'just get the thing in the air and we can fly it.'

The combination of all these factors eventually led to the tragedy on Pad 34.

However, the lessons were learned. The command module underwent changes. The number of flammable components within the interior was drastically reduced and more stringent tests and construction checks were carried out. A less complex outward opening hatch was designed and installed along with other details in what amounted to a total design review. The pure oxygen ground atmosphere was replaced with a nitrogen oxygen mixture that gradually switched to pure oxygen after launch. Confidence began to return when the astronauts started to say in public that they were delighted with the changes. In answer to a question posed by a congressman Frank Borman answered:

> You are asking us do we have confidence in the spacecraft, NASA management, our own training, and… our leaders. I am almost embarrassed because our answers appear to be a party line. Everything I said last week has been repeated by the people I see here today. The response we have given is the same because it is the truth…. We are trying to tell you that we are confident in our management, and in our engineering and in ourselves. I think the question is really: Are you confident in us?

It is arguable that without the fire the command module would never have been safe to fly and that a more serious disaster would have eventually taken place. Certainly, in the rush to get to the Moon, a pause was needed. As Grissom said some weeks before the fire:

> We're in a risky business … and we hope if anything happens to us, it will not delay the program. The conquest of space is worth the risk of life…. Our God-given curiosity will force us to go there ourselves because in the final analysis only man can fully evaluate the moon in terms understandable to other men.

Sentiments that apply equally, in my view, today.

7.2 TESTING PHASE: APOLLOS 4–6

No missions were designated Apollo 2 or 3. To honour the widows' wishes, the crew of the fatal fire retained the Apollo 1 mission number. As there had been three Apollo/Saturn Ib unmanned flights prior to the tragedy, the numbering picked up with Apollo 4.

7.2.1 APOLLO 4 LAUNCH (NOVEMBER 4, 1967)

The Apollo 4 mission was the first launch of the Saturn V booster and carried a command and service module (CSM) combination into a 190 km Earth orbit (Figure 7.1). After two orbits, the third stage fired again (its first re-ignition in space) to push the craft into an elliptical orbit. This orbit was then trimmed after the CSM separated from the third stage and fired the service propulsion system (SPS). The final orbit had an apogee of 18,092 km and a perigee of −74 m, below the Earth's surface! This seemingly odd choice ensured that the command module would re-enter Earth's atmosphere at high speed (the third stage orbit also had a sub-surface perigee ensuring that it would burn up safely in the atmosphere). Shortly after the CSM passed orbital apogee, the SPS was remotely fired again to accelerate the craft to a speed comparable to that experienced on the return from a lunar mission. The command module separated and re-entered successfully.

7.2.2 APOLLO 5 LAUNCH (JANUARY 22, 1968)

This was the first launch of the lunar module (designated LM-1) which flew without a CSM stack. Ground control tested the descent and ascent engine systems, and in the process, the descent engine became the first ever throttleable engine to be fired in space. During the mission, the landing abort procedure (firing the ascent stage while still attached to the descent stage) was tested. As a result of a successful sequence of tests, a further unmanned launch of LM-2 was cancelled and LM-3 was cleared for use in the Apollo 9

Figure 7.1 The launch of Apollo 4. (Image courtesy of NASA.)

flight. Subsequently, LM-2 was used for ground testing and is on display in the Boeing Milestones of Flight Hall at the National Air and Space Museum in Washington, DC.

7.2.3 APOLLO 6 LAUNCH (APRIL 4, 1968)

For this test flight, a Saturn V booster lifted a simulated payload corresponding to ~80% of a full-up Apollo mission. The third stage was used to fire a CSM and a simulated LM into a translunar trajectory. A direct return abort was then carried out by firing the SPS, resulting in a flight time of 10 hours. After that burn, the SPS was used to increase the entry speed to something nearer to that after a lunar mission return. After another successful re-entry, and as Apollo 4 had already tested the third stage re-start, a further unmanned flight was cancelled and the Saturn V approved for a manned launch.

The rocket, however, had not performed faultlessly. During the flight, longitudinal vibrations (*pogo oscillations*) set in. They arose from variations in the combustion rate, which in turn affected the propellant pressure and reinforced the oscillations. The vibrations became so severe that some of the J-2 engine fuel lines in the second and third stages ruptured and caused the engines to shut down. The onboard guidance successfully compensated by burning the second and third stages for longer, although the resulting orbit was more elliptical than had been planned. Despite this, the overall success of the systems tested gave NASA enough confidence to move on.

7.3 THE FORGOTTEN MISSION: APOLLO 7 (OCTOBER 11–22, 1968)

Call Sign: Apollo 7

On October 11, 1968, Wally Schirra (MC, Mercury 5, Gemini 6), Donn Eisele (CMP) and Walt Cunningham (LMP) lifted off for an 11-day Earth orbital test of the Apollo command and service module combination. This mission has in many ways been somewhat eclipsed by the more historic achievements of the ones that followed, which is a shame. As the first manned flight of the Apollo programme and the first time the United States ventured into space since the Apollo 1 fire, NASA had a lot resting on the mission from the technical and political standpoints.

After one and a half orbits, Schirra tested the manoeuvrability of the craft by separating from the Saturn Ib third stage and simulating a docking, as a rehearsal for the CSM–LM linking that would be required on Moon missions. The SPS successfully fired eight times, on the first occasion surprising the crew with the jolt it conveyed. Although hardly an engineering highlight, the mission also included the first-ever live TV transmission from space.

The mission was a total success and the Apollo spacecraft performed almost flawlessly for 11 days, longer than required for a complete Moon mission. Without that major achievement, the Apollo program and Apollo 8 in particular would have been held back.

7.4 THE MOST DARING MISSION: APOLLO 8 (DECEMBER 21–27, 1968)

Call Sign: Apollo 8

The period between 1967 and 1968 was not the most encouraging for NASA. The aftermath of the Apollo 1 fire led many to question the concept of an attempted Moon landing. The Saturn V rocket flew, but not without 'teething problems' and lunar module development did not proceed as quickly as expected. In the background, the Vietnam War rumbled on, consuming American resources and morale,[4] which was already badly sapped by the assassinations of Martin Luther King and Robert Kennedy.

As 1968 drew to a close, the Apollo 7 CSM stack sat atop a Saturn Ib rocket ready for its October flight. Apollo 8 was scheduled to be the first manned mission with the lunar module. However, delays in delivering a working LM to NASA probably meant postponing that mission until the New Year, putting the end-of-the-decade target for the landing in doubt.

As August approached, the idea for a daring re-tasking of the Apollo 8 mission occurred to George Low, the NASA engineer responsible for overseeing the development of the Apollo spacecraft. If Apollo 7 flew well and if the LM would not be ready for the planned Apollo 8, then the time could be used for a mission to the Moon. The Saturn V booster would need to fly with astronauts onboard at some point. Since it was capable of flinging the CSM to the Moon, it seemed wasteful to use it simply for an Earth orbit mission.

Predictably, the early reaction among NASA management was not enthusiastic. James Webb, the NASA administrator, was at the US Embassy in Vienna when he heard the idea. Apparently he yelled, 'Are you out of your mind?' over a transatlantic phone line. A Moon flight was certainly a bold idea. For one thing guidance systems (including software) for a lunar mission had not yet been fully worked out. Also, sending the second manned Apollo flight out to the Moon without a lunar module (considered a viable lifeboat in case of an emergency) could jeopardise the whole NASA organisation.

However, another factor came into play. According to the CIA, the Soviets were about to resume launches of Soyuz spacecraft after the disastrous flight of Soyuz 1, culminating in the death of Vladimir Komarov when his craft crashed into the ground after re-entry. Although the American intelligence agencies were reasonably sure that the Soviets could not land a man on the Moon before the end of the decade (they had not tested any sufficiently powerful boosters), they were capable of sending a spaceship on a loop around the Moon. If they accomplished that, they could reasonably claim to have beaten NASA to the Moon.

In fact, as we now know, the Soviets were planning an ambitious program called *Zond* that involved flying several tanks of fuel into Earth orbit where they would be combined with a small manned craft for a landing on the Moon. The plan faltered after Komarov's death and was re-structured as an attempt to orbit the Moon before the Americans. In September 1968, the Soviets flew an unmanned Zond around the Moon and 2 months later, another craft loaded with tortoises, flies and worms took the same trip. Then, for once, the Russians faltered. They wanted more tests. The Cosmonauts wanted to proceed but were held back. The United States might still fail; it was better not to risk all so soon.

Within the planning teams at NASA the response to the prospect of sending a crew to orbit the Moon was more enthusiastic. James Webb understood the potential political payoff of a successful mission, both abroad and at home, and, on reflection, agreed to give the idea careful consideration. After all, the US Congress needed to see a major NASA success to ensure the continuity of funding. When the engineering staff, including Wernher von Braun and the Marshall Space Flight Center team, said they could be ready to fly to the Moon by December, Webb sanctioned the plan, provided Apollo 7 flew without any problems.

Sixty-six hours after liftoff, Apollo 8 slipped behind the Moon on its free return trajectory. The crew, Frank Borman (MC, Gemini 7), Jim Lovell (CMP, Gemini 7, Gemini 12) and Bill Anders (LMP), readied themselves and their ship for the crucial SPS burn that would slow them into lunar orbit. In the process, they became the first people to see the far side of the Moon with the naked eye. Theirs was a mission of firsts. On the way, the Saturn V third stage boosted them to a speed faster than any human had ever experienced. They were also the first people to view the Earth from such a distance that the entire globe could be seen at once (Figure 7.2).

Christmas Eve 1968 saw the most extraordinary TV broadcast in the history of mankind. The crew of Apollo 8 caught Earthrise above the Moon on camera for all to see (see Figure 2.1). Historically, it was a moment of supreme significance. The United States was involved in a bloody war that many of its citizens

Figure 7.2 One of the first photographs, probably taken by Bill Anders, showing the whole Earth. This photo from Apollo 8 was taken from a distance of about 30,000 km. South America appears in the top half centre. Africa is entering into the shadow (bottom left); North America is at the bottom right. (Image courtesy of NASA.)

considered immoral. Civil rights leaders struggled to come to terms with the death of their icon, and people were still astounded by the assaults on democracy implied by the assassination of a president. In the midst of this black year, the crew of Apollo 8 showed the world a view of the Earth with no political, racial or geographical boundaries. Furthermore, the astronauts chose to end their broadcast by reciting lines from the Book of Genesis: 'In the beginning God created the heaven and the Earth; and the Earth was without form and void...' It is impossible to convey the impact that this had on the viewers (see Chapter 2). The broadcast became the most watched TV program to that date.

After 10 orbits lasting 20 hours, the next milestone was to fire the SPS in order to inject the crew back onto a free return path. The burn had to be carried out behind the Moon. The tension in the crew was evident to those who listened:

Lovell: Did you guys ever think that one Christmas Eve you'd be orbiting the Moon?
Anders: Just hope we're not doing it on New Year's.

After the burn succeeded and the spacecraft emerged from behind the Moon, Lovell had the following exchange with the CAPCOM[5] (Ken Mattingly):

Lovell: Houston, Apollo 8. Over.
Mattingly: Hello Apollo 8. Loud and Clear.
Lovell: Please be informed there is a Santa Claus.
Mattingly: That's affirmative. You are the best ones to know.

As it transpired, the lasting achievement of NASA's most daring mission was one that no one expected: a new perspective on the Earth. Many people date the start of the environmental movement to the first sight of the Earth rising over the lunar highlands. In the words of Bill Anders, 'we flew all that way to explore the Moon, and the most important thing we discovered was the Earth.'

7.5 THE LM FLIES: APOLLO 9 (MARCH 3–13, 1969)

Call Signs: CM: *Gumdrop*; LM: *Spider*
With the lunar module finally ready for manned spaceflight testing, the Apollo 9 mission faced an ambitious series of targets. At just over 2 hours and 40 minutes after the Saturn V launch, Jim McDivitt (MC, Gemini 4), Dave Scott (CMP, Gemini G8) and Rusty Schweickart (LMP) were ready for the opening milestone: the docking and extraction of the lunar module.

Dave Scott was the first astronaut to practice this vital manoeuvre outside the mission simulator. The docking mechanism worked perfectly and the LM was withdrawn from its protective housing at the top of the third stage (Figure 1.2). After checking the connecting tunnel, the SPS was fired to demonstrate that the docked ships could be manoeuvred by one engine. The third stage engine was then remotely fired to move it into a solar orbit, out of the way. Two more SPS firings adjusted the crew's orbit and proved the stack's stability under engine gimballing. By this time, the mission was entering its third day and McDivitt and Schweickart crossed to the lunar module (the first internal crew transfer between spacecraft).

With the hatches closed, the LM's independent environmental systems were checked. A successful firing of the descent stage engine demonstrated that it could act as a reliable backup in case of SPS failure (which proved crucial during Apollo 13). However, despite all these successes, another factor threatened to put the mission plans in jeopardy. The contortions required to don spacesuits before moving into the LM induced a form of motion sickness in both McDivitt and Schweickart, who actually vomited twice. Mission rules called for Schweickart to perform an extravehicular activity (EVA) with the lunar surface backpack (the PLSS described in Section 6.4.1) before the craft could be separated. It was vital to show that the crew could transfer back to the CM via space should re-docking be impossible. If Schweickart were to vomit in his spacesuit helmet the consequences could have been lethal, so McDivitt cancelled the EVA scheduled for mission day 4. Fortunately, by that stage of the 4th day, Schweickart felt considerably better and McDivitt allowed the EVA to take place. Schweickart stood on the lunar module's porch and retrieved some experiments from the LM's exterior while Scott participated in the EVA from the command module, retrieving experiments and standing by to assist if necessary (Figure 7.3).

Figure 7.3 Dave Scott standing in the Apollo 9 command module hatch. The photo was taken by Rusty Schweickart from the LM porch during their joint EVA. (Image courtesy of NASA.)

On day 5, McDivitt and Schweickart entered the LM once again and Scott operated the release mechanism to separate the two craft. After 45 minutes of loitering near to Gumdrop, Spider's descent engine was fired to move the craft onto a slightly higher orbit so that the command module would pull away over time. The day then continued with various tests of the descent engine, culminating in a burn to lower the orbit so that the LM would catch up with the CM. By the time of the burn, the two craft were separated by 185 km. The descent stage was jettisoned, and the ascent engine fired to simulate a return from the surface. McDivitt brought the LM back and docked with the command module, demonstrating that the LM was capable of performing the task even though on the lunar missions the final manoeuvres over the last few metres would be handled by the command module. This concluded the first manned flight of a spacecraft that was not capable of re-entry. The LM-3 ascent stage was finally jettisoned, and the ascent engine fired remotely as a final test.

7.6 THE LM'S FIRST JOURNEY TO THE MOON: APOLLO 10 (MAY 18–26, 1969)

Call Signs: CM: *Charlie Brown*; LM: *Snoopy*

As the first full dress rehearsal for a Moon landing, Apollo 10 was another highly significant mission. Tom Stafford (MC, Gemini 6, Gemini 9), John Young (CMP, Gemini 3, Gemini 10) and Eugene Cernan (LMP, Gemini 9) dry-ran all stages of the landing up to the initiation of the powered descent designed to take the LM to the surface. Docking and extraction of the LM took place after translunar injection and the first separation of the LM and CSM in lunar orbit happened 3 days later.

Young then became the first astronaut to fly the command module solo in lunar orbit. Stafford and Cernan used the descent stage engine to put Snoopy into the descent orbit and tested systems including the landing radar as they dropped to ~15 km altitude at which the powered descent would be initiated on Apollo 11. This also gave them the opportunity to survey the proposed landing site.

After separation from the descent stage, the plan was to test the abort procedures by getting the onboard computer to fire the ascent engine and return them to the CSM using the abort program. However, just after separation and before the crew had time to initiate the abort procedure, the ascent stage snapped into a violent tumbling roll. Cernan later recalled seeing the lunar horizon sweep past 8 times in 15 s. Stafford acted quickly, shutting systems down and initiating manual control. He was able to stabilise the ship even though the roll was close to the rate at which the crew members would have blacked out.

As it transpired the problem arose from a simple human error. In order to set up the abort program, a mode switch had to be set. Stafford, following the mission procedure cards, flipped the switch correctly and then Cernan reached over from his station and, with the instinct of a training pilot, flipped the switch without looking at it, returning the mode to its initial setting. As a result, the computer was not in abort mode and, without a radar lock on the CSM, set about hunting for the ship by rotating the ascent stage. A few more rotations would have bled off too much energy, causing the ship to crash out of orbit The rest of the return to docking with the CSM went without a hitch (Figure 7.4).

Figure 7.4 The ascent stage of the Apollo 10 lunar module (*Snoopy*) shortly before redocking with the CSM in lunar orbit. (Image courtesy of NASA and John Young.)

Amusingly, the ascent engine aboard *Snoopy* was fuelled only for the mission's planned manoeuvres. This mitigated against the crew 'going rogue' and deciding to slip to the head of the line by taking the LM down to land while they were close by. Had they done so, there would not have been enough fuel for the ascent engine to return them to orbit.

Apollo 10 clocked up a number of records. During the return flight, the ship reached the highest speed recorded by a manned vessel at 11.08 km/s. The crew also set the record for being the humans who have travelled furthest from home at 408,950 km. Although all the lunar missions sat at 111 km above the lunar surface, the Earth-to-Moon distance varies 43,000 km over the course of the Moon's orbit. During Apollo 10, the Moon was slightly nearer to Earth than on other missions, but the mission reached the furthest distance of its orbit on the far side of the Moon as the Earth's rotation took Houston almost a full Earth diameter away. As a result, the astronauts were further away from their homes and families than any other crew.

7.7 THE LANDING: APOLLO 11 (JULY 16–24, 1969)

Call Signs: CM: *Columbia*; LM: *Eagle*
Time on surface: 21 hours 36 mins
An estimated 1 million spectators watched the launch of Apollo 11, crowding onto the roads and beaches in the vicinity of the launch site. In addition, a substantial press corps was present to document the historic occasion (Figure 7.5). Neil Armstrong (MC, Gemini 8), Mike Collins (CMP, Gemini 10) and Buzz Aldrin (LMP, Gemini 12) took off on July 16, 1969 at 13:32:00 UTC.

The Apollo 11 landing site was selected after considering several factors including:

- the need for a smooth area with as little cratering as possible;
- having an approach trajectory without tall hills or deep craters that might throw out the landing radar;
- accessibility from the lunar orbit resulting from a free-return trajectory;
- good visibility for landing: Sun between 7 and 20 degrees behind the lunar module on approach:

Figure 7.5 The Apollo 11 launch as seen from the press area. (Image courtesy of NASA.)

although the last factor had more to do with determining the date of launch. The timing of the landing, just after lunar dawn, was selected to minimise the range of temperatures the crew would experience on the surface. The Apollo 10 LM flew within 15 km of the selected site and confirmed its suitability.

The events of the landing itself have already been discussed in Chapter 1. The mission plan called for the astronauts to settle down for 5 hours of sleep after the post-landing checks were completed and prior to their first and only EVA. However, Armstrong and Aldrin, with NASA's approval, chose to get ready early, understandably thinking that they would not be able to sleep.

In fact, starting early turned out to be a good idea. In the cramped confines of the LM and with quite a lot of additional detritus such as checklists and other items around, the preparations took a full 90 minutes longer than they had during training. Having depressurised the cabin, the crew opened the forward hatch. Armstrong had difficulty manoeuvring through the hatch with the PLSS backpack on, so it was some 11 minutes after hatch opening before he started to descend the ladder. On the way, he pulled on a D-ring cable to deploy an equipment stowage assembly from the descent stage and turn on a TV camera to relay the historic first step on the Moon.

The image of Armstrong stepping off the LM has become iconic. However, that image and the later sequences when Aldrin joined him (including the telephone call from President Nixon) are blurry and 'ghostly'. The camera on the Moon was capable of reasonably high-quality transmission but the signals were not compatible with commercial TV technology in 1969. The images were received at various tracking stations and then passed onto the global audience by pointing conventional TV cameras at monitors displaying the live feed.

In 2009, NASA released results from a restoration project run by a team of retired NASA employees and contractors who tried to find the magnetic tape recordings of the original raw signals. Unfortunately, the recordings had been erased, but the group found videotapes and a Super 8 movie of one of the monitors that allowed them to show some enhanced sequences. Lowry Digital completed a full restoration in late 2009.[6]

In total, the first Moonwalk lasted nearly 2 hours, 32 minutes, during which time samples were collected, a passive seismometer and laser reflector (to allow accurate determination of the Earth-to-Moon distance by bouncing laser beams from Earth) was deployed and a US flag planted. The astronauts also left an Apollo 1 mission patch on the surface.

Preparations for the return to lunar orbit did not pass without incident. While moving about the cabin, Aldrin accidentally damaged the push switch that would arm the ascent engine. The plastic knob on the switch broke off. While the astronauts slept, the mission control team found a way to bypass the switch and use other circuits to achieve the same end. In fact, on the Moon, the astronauts resolved the problem by jamming a pen into the hole. On future flights, guards were installed around certain controls to prevent such breakages from happening again.

During the activity on the surface, Collins circled the Moon in the command module. The first time he travelled behind the Moon while the LM was on the side facing the Earth, he was cut off from human contact in a way that no individual had ever experienced before. As he puts it in his book[7]:

> I don't mean to deny a feeling of solitude. It is there, reinforced by the fact that radio contact with the Earth abruptly cuts off at the instant I disappear behind the Moon. I am alone now, truly alone, and absolutely isolated from any known life. I am it. If a count were taken, the score would be three billion plus two over on the other side of the Moon, and one plus God only knows what on this side. I feel this powerfully – not as fear or loneliness – but as awareness, anticipation, satisfaction, confidence, almost exultation. I like the feeling. Outside my window, I can see stars – and that is all. Where I know the Moon to be, there is simply a black void; the Moon's presence is defined solely by the absence of stars.

Docking the ascent stage with the command module, its subsequent jettison to crash on the Moon, firing of the SPS to return to Earth and eventual splashdown all occurred without incident.

Almost immediately after the completion of the Apollo 11 mission, public interest in the space program began to decline. There was a brief re-surfacing of popular engagement during the hazardous Apollo 13 flight, but the tide had turned. After a period of guaranteed presidential support, NASA was now working in a different climate. Apollo was not the only program NASA wished to develop. The Mariner probes (1962–1973) visited Venus and Mars, and transmitted spectacular photographs demonstrating that manned exploration was not the only way to obtain scientific data from space. Planning for the Viking mission (1975) to soft-land a probe on Mars, the *Skylab* Space Station (1973) and the early development of the space shuttle, were all taking place during the Apollo years and NASA had to provide funds for these as well.

The problem NASA faced was the perception that Apollo was staged primarily as a race against the Soviets. The scientific benefits of the program had not been stressed. Now that the race had been won, the science had to move to the forefront to justify more flights.

In 1970, NASA requested a $3.33 billion budget for 1971 – a $500 million cut from the previous year and suggested that Apollo 20 be cancelled. In the end, NASA had to settle for about $3.27 billion. More hard decisions followed. On September 2, 1970, NASA announced the cancellation of two more Apollo flights, which triggered an uproar in the press. The *New York Times* noted that the savings from cancelling three Apollo flights amounted to no more than 2% of the NASA budget and 0.25% of the total invested in the Apollo programme to date: 'This decision can only vindicate the critics who have insisted that Apollo was motivated by purely prestige considerations, not scientific goals.'

Each Apollo flight represented a cost of $20 million compared with the $25 billion already spent. All the spacecraft and rocket boosters had been built. (In the end one of the rockets was used to launch *Skylab* and the other two are on display at the Kennedy and Johnson Space Flight Centers.) However, NASA wanted to support a post-Apollo program and there was no valid alternative to dropping the missions.

7.8 PRECISION LANDING AND ALL-WEATHER TESTING: APOLLO 12 (NOVEMBER 14–24, 1969)

Call Signs: CM: *Yankee Clipper*; LM: *Intrepid*
Time on surface: 1 day 7 hours
Although the problems surrounding the Apollo 11 landing were not widely known outside NASA, pride dictated that the next landing would be more 'on-target', especially as the geologists planning for the later missions wanted to be sure that the astronauts would collect rocks from certain specified areas. Some sources of error in the original landing plan were sorted out and two changes were made to the software. The first allowed the crew to fine-tune the optical references of the Landing Point Designator (LPD) by calibrating against a star during powered descent. The second enabled the crew to update guidance information.

As the LM emerged from behind the Moon on the last orbit prior to powered descent, the ground radar checked the craft's velocity and compared it to the ideal value. This enabled mission control to calculate the difference between where they wanted to land and where the computer (unaware of any velocity deviation) would guide the spacecraft. This information was radioed to the crew who entered it into their computer. As altering the data from the navigation platform would have been too complicated, the new information was incorporated by shifting the designated landing spot instead. The software treated the landing target as a point relative to the LM's position, rather than a location on the Moon, in any case.

NASA decided that Apollo 12 would attempt a pin-point landing within walking distance of the *Surveyor 3* probe that soft-landed on the Moon on April 20, 1967. The plan was to determine the effects of long-term exposure to space on the probe. The site was also of geological interest, so the two objectives dovetailed nicely.

Perhaps in honour of the all Navy crew, Pete Conrad (MC, Gemini 5, Gemini 11), Dick Gordon (CMP, Gemini 11) and Al Bean (LMP), Apollo 12's launch took place in a rainstrom. In itself, this presented no problems but the Saturn V booster was struck by two lightning discharges, the first of which followed the ion trail down the exhaust plume to the launch tower and knocked out all three fuel cells in the service module. The second took down the CSM's guidance system. With a completely garbled telemetry stream pouring into mission control and virtually every warning light and alarm going off in the command module, the flight appeared moments away from being aborted. However, quick thinking on the ground circumvented that possibility.

John Aaron, the Electrical, Environmental and Consumables Manager (EECOM), realised that he had seen a similar set of muddled telemetry during an earlier test when a power supply failure in a CSM impacted the data. Aaron called the flight director on the mission control communications loop: 'Flight, EECOM. Try SCE to Aux.' The command was relayed to the crew, but neither the flight director nor the CAPCOM nor Pete Conrad recognised the relevant switch (Figure 7.6).

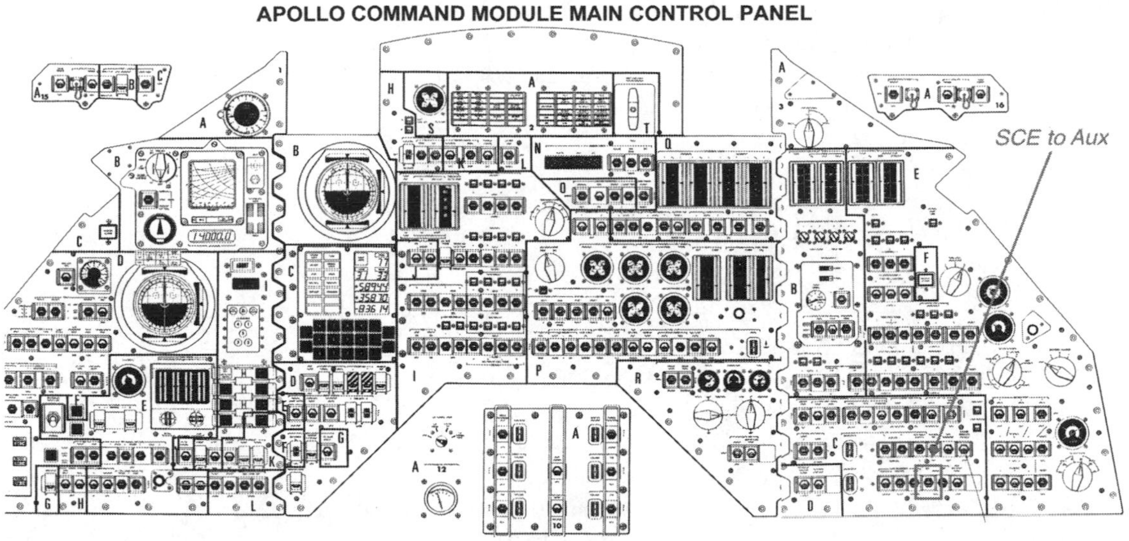

APOLLO COMMAND MODULE MAIN CONTROL PANEL

SCE to Aux

Figure 7.6 The location of the SCE mode switch on the command module main control panel. (Image courtesy of NASA with the author's annotation.)

Fortunately, Al Bean had come across that control during a training incident a year earlier when the solution to a simulated problem required the same setting change. The Signal Conditioning Electronics (SCE) system would operate correctly in a low-voltage situation if switched to the aux position. Once this had been done, telemetry from Apollo 12 was restored.[8]

Meanwhile, the Saturn V continued to fly correctly since its guidance systems were unaffected by the lightning. The fuel cells were put back on line, and once the astronauts had settled into orbit, they checked out the ship and re-aligned the CSM navigation platform. The only remaining worry was that the strike might have prematurely triggered the pyrotechnics that deployed the parachutes during re-entry. As there was no way to check this, and certainly no way to repair any issues, mission control decided not to tell the astronauts and allowed the crew to continue to the Moon. Without the parachutes, they could not survive a re-entry from Earth's orbit, never mind a return from the Moon, so there was nothing to lose by letting the mission continue.

The mission transcript for this period is remarkable. (Astronaut Gerald Carr was the mission CAPCOM. He later commanded the *Skylab* Space Station on the third and final mission to that facility.)

From the first indication of a possible issue at the 37-s point:

000:00:37 Gordon (onboard): What the hell was that?
000:00:38 Conrad (onboard): Huh?
000:00:39 Gordon (onboard): I lost a whole bunch of stuff; I don't know…
000:00:40 Conrad (onboard): Turn off the buses.
Public Affairs Office: 40 seconds.
000:00:42 Carr: Mark.
000:00:43 Carr: One Bravo.
000:00:43 Conrad (onboard): Roger. We had a whole bunch of buses drop out.
000:00:44 Conrad: Roger. We [Garble] on that. [Long pause.]
000:00:45 Bean (onboard): There's nothing; it's nothing…
000:00:47 Gordon (onboard): A circuit…
000:00:48 Conrad (onboard): Where are we going?
000:00:50 Gordon (onboard): I can't see; there's something wrong.
000:00:51 Conrad (onboard): AC Bus 1 light, all the fuel cells…
000:00:56 Conrad (onboard): I just lost the platform.
[The platform was the CSM guidance system.]

Public Affairs Office: Altitude a mile and a half now; velocity 1,592 feet per second.
000:01:00 Bean: [Garble] Got your GDC.
000:01:02 Conrad: Okay, we just lost the platform, gang. I don't know what happened here; we had everything in the world drop out.
000:01:08 Carr: Roger.
Public Affairs Office: Plus one.
000:01:09 Gordon (onboard): I can't. There's nothing I can tell is wrong, Pete.
000:01:12 Conrad: I got three fuel cell lights, an AC bus light, a fuel cell disconnect, AC bus overload 1 and 2, main bus A and B out. [Long pause.]

to the instruction from EECOM:

000:01:36 Carr: Apollo 12, Houston. Try SCE to auxiliary. Over.
000:01:39 Conrad: Try FCE to Auxiliary. What the hell is that?
000:01:41 Conrad: NCE to auxiliary...
000:01:42 Gordon (onboard): Fuel cell...
000:01:43 Carr: SCE, SCE to auxiliary. [Long pause.]
000:01:45 Conrad (onboard): Try the buses. Get the buses back on the line.
000:01:48 Bean (onboard): It looks – Everything looks good.
000:01:50 Conrad (onboard): SCE to Aux.
000:01:52 Gordon (onboard): The GDC is good.

took only 59 seconds. Then came Conrad's wry remark:

000:03:17 Conrad: Think we need to do a little more all-weather testing.

A true feel for the nature of this crew comes out a little later in the transcript:

000:06:43 Gordon (onboard): Man, oh man...
000:06:44 Bean (onboard): Isn't that a...
000:06:45 Conrad (onboard): Wasn't that a SIM they ever gave us?
000:06:46 Gordon (onboard): Jesus!
000:06:50 Conrad (onboard): [Laughter.]
000:06:51 Gordon (onboard): That was something else. I never saw so many...
000:06:52 Conrad (onboard): [Laughter.]
000:06:54 Gordon (onboard): There were so many lights up there, I couldn't even read them all.
000:06:55 Conrad (onboard): [Laughter.]
000:06:57 Gordon (onboard): There was no sense reading them because there was, I was, I was looking at this; Al was looking over there.
000:07:02 Conrad (onboard): Everything looked great [Laughter] except we had all the lights on...

The lunar landing was made using the new guidance system which only needed manual correction in the last phase of descent. During approach, the computer displayed data that allowed the commander to pick out the landing point targeted by the guidance system.

Two sets of vertical lines with horizontal calibrations were ruled on the commander's window. He aligned himself so that the vertical lines were superimposed and then looked down the horizontal calibrations to the angle designated by the computer. As part of the process, the computer yawed the LM slightly from left to right effectively 'drawing' a horizontal line across the surface to help the commander visualise the target area. As the computer was still flying the landing at this stage, any corrections made by the commander on his controls effectively moved the designated landing site instead of the attitude of the ship. With the Apollo 11 landing, Armstrong had found the system hard to use as the sloshing fuel made the craft wobble.

The amount of dust thrown up by the descent engines was far greater than during the Apollo 11 landing. As a result, Pete Conrad could see little through the windows from about 30 m down. Nevertheless, the touchdown was very smooth. Within moments of stepping on the surface, Conrad could see *Surveyor 3* about 160 m away on the side of a crater (Figures 7.7 and 7.8).

The crew performed two EVAs lasting 7 hours, 45 minutes. They collected lunar samples and pieces of the *Surveyor* probe to return to Earth for analysis as well as setting up experiments on the surface.

Figure 7.7 Pete Conrad next to *Surveyor 3* with the Apollo 12 lunar module in the background. (Image courtesy of NASA and Al Bean.)

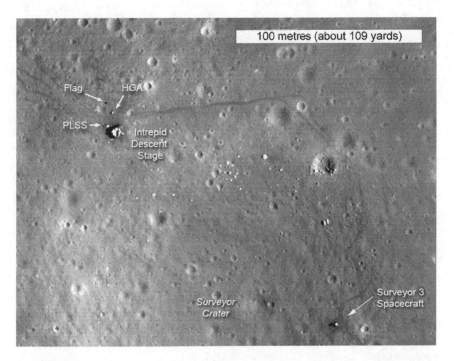

Figure 7.8 The Apollo 12 landing site in *Oceanus Procellarum* imaged during the second Lunar Reconnaissance Orbiter (LRO) low-altitude campaign. (Image courtesy of NASA, Goddard Space Flight Center and Arizona State University.)

7.9 THE ALMOST TRAGEDY: APOLLO 13 (APRIL 11–17, 1970)

Call Signs: CM: *Odyssey*; LM: *Aquarius*

The Apollo 13 mission in many ways represented NASA's finest hour during the Moon exploration program. At one level, the mission was a failure since no lunar orbit or landing occurred. An explosion deprived Jim Lovell (MC, Apollo 8), Jack Swigert (CMP) and Fred Haise (LMP) of electrical power and oxygen putting them in mortal danger. However, considering the brilliant improvisation that went on, the endurance (both mental and physical) shown by crew and mission controllers, the way in which various systems were pushed to their design limits and functioned perfectly and the tremendous support expressed by people around the world, the mission has to be seen as a hugely positive one for the NASA community.

The story of the Apollo 13 accident starts before completion of the Gemini program. North American Rockwell, the command and service module contractor, subcontracted the construction of the oxygen and hydrogen tanks, used to store the liquids needed to feed the fuel cells, to Beech Aircraft. The contents of the tanks were normally kept at −207°C, at which temperature they had a rather slushy consistency. However, just enough of the oxygen and hydrogen would evaporate off to feed the pipes running to the fuel cells and in the case of the oxygen, to the command module cabin. Occasionally, the tank pressure dropped, and so heating and stirring systems were incorporated into the tanks.

Early in the service module design, its operating voltage was fixed at the 28 V supplied by the fuel cells. This information was passed on to Beech and its engineers designed the heaters and thermostats to operate at 28 V. However, while sitting on the launch pad, the service module would run on external power at 65 V. When North American became concerned that the difference would cause problems in the tanks, it instructed Beech to up-rate the components. However, for some reason the thermostat switches were never replaced with the 65 V versions. NASA, North American and Beech all reviewed the designs but the mistake was not caught. All the Apollo service modules up to 13 flew with thermostat switches rated at only 28 V.

The next event in the chain leading to explosion occurred when the tanks were delivered to North American for installation. The vessel that eventually became oxygen tank 2 on Apollo 13 started its life in the Apollo 10 service module. However, due to a re-design, the tank was removed from that module and replaced by a newer version. The extracted tank was modified and then installed on the service module destined for Apollo 13. During removal, one of the bolts holding the sector 4 shelf on which the tank was mounted was not removed. A crane tried to lift the tank shelf, which moved upward and then fell back with a small thump. The tanks on the shelf were inspected for faults and found to be fine.

Unfortunately, a problem with the oxygen tank was not discovered until the countdown test. During this procedure, all propellants and other fluids were loaded into the spacecraft on the launch pad and the crew was stationed in the command module for a full rehearsal. Once the test was complete, the tanks were emptied. In the case of the oxygen tanks in the SM, this was done by pumping pressurised oxygen to force the liquid out via the drain pipes. Unfortunately, tank 2 was not emptying. The engineers studied the problem for 11 days and eventually surmised that one of the drain lines had been jolted by the attempted tank lift so that the pressurised oxygen leaked from the inlet directly to the drain without forcing the liquid anywhere.

Annoying though this was, the mission was not threatened. The drain line would never be used in flight. As long as the tank could be emptied in the meantime, there would not be a problem on launch day. Considering the time required to remove the faulty tank, replace it and check out a new one, the decision was made not to risk a delay to the launch and switch on the heater in the tank. Boiling off the liquid oxygen was a good way of getting it out of the tank.

However, no one realised that the thermostats, designed to turn off the heaters at 27°C, were rated only for 28 V. Estimating that 8 hours were needed to boil off the liquid, the engineers switched on the heater in tank 2 and ran it from the external power supply. When the time came for the thermostat switches to open, the higher current driven by the 65 V line fused them shut. The temperature in tank 2 rose to ∼500°C, but the ground technicians were unaware of the increase. As the design called for the thermostats to cut power to the heaters at 27°C, the scale on the gauges did not run any higher. Inside tank 2, the Teflon insulation around the cables leading to the fan motor melted.

Seventeen days later and 322,000 km into space, Jack Swigert, the command module pilot, threw the switch to stir the contents of oxygen tank 2. Earlier in the mission, the tank had been full enough for the slushy liquid to cover the wires; now they were exposed. A spark flew igniting the gaseous oxygen and the sudden build-up of pressure blew the top off the tank. Now exposed to a vacuum, the contents of tank 2 evaporated in a flash filling sector 4 of the module and forcing the side of the sector clear away from the spacecraft (Figure 7.9).

The other oxygen tank was undamaged, but because it shared some pipe work with tank 2, it also started to leak as its contents found a path into space. The explosion jammed some of the valves running to the thrusters. When the stack was rocked by the rupture of tank 2 and the subsequent venting, the autopilot started firing the thrusters to compensate. Some failed to fire, so the autopilot started hunting for a combination that would steady the ship.

With the fuel cells fed by tank 2 dead and the other dying and with the oxygen needed to breathe in the command module visibly venting into space, the crew faced an imminently fatal situation. The solution was to transfer to the lunar module and live there for the remainder of the flight. In a frantic few minutes, the crew had to power up the LM, transfer guidance control from the CM to the LM and shut down all the electrical systems in the CM in an effort to preserve the batteries needed to power the craft during re-entry.

The rest of the mission became a story of survival. The LM descent engine was used to provide the ΔV needed for the transfer to a free return trajectory. With all electrical power turned off, the cabin temperature dropped to the level of a meat storage freezer and the drinking water froze. Mission control instructed the crew to stop dumping their urine overboard because the crystalline cloud it formed was making it harder to track the spacecraft.

Figure 7.9 The Apollo 13 service module after jettison 4 hours before re-entry. The damaged section can be seen clearly. (Image courtesy of NASA.)

Figure 7.10 The cut-and-paste fix to the lunar module's CO_2 filtering system during the Apollo 13 mission. (Image courtesy of NASA; scan by John Fongheiser.)

With all three astronauts living in the LM, CO_2 levels built up dangerously since the LM's environmental control system was designed to deal with the respiration of only two people. The lithium hydroxide filters used to remove CO_2 from the LM were becoming saturated. While there were plenty of filters for the command module, they were the wrong shape to fit in the LM. A team in mission control found a somewhat primitive way to adapt the filters by using a hose from the spacesuits, the covers from the flight plan, lengths of tape and various other items scrounged from the spacecraft. Tense moments were spent building this kludge, following instructions radioed from the ground, but the system worked (Figure 7.10).

The complete story of Apollo 13 was immortalised in the film of the same name and detailed in the book Jim Lovell wrote with Jeffrey Kluger.[9]

After the accident investigation report was issued, alterations were made to the spacecraft design to prevent such occurrences in the future. Apollo 14 was modified to include an additional oxygen tank, a backup battery in the service module and elimination of the fans and thermostat switches from the oxygen tanks. Storage for an emergency water supply was added to the command module.

After the safe return of the crew, Grumman sent an invoice (in jest) to North American for towing the CSM (Figure 7.11). Apparently, North American did not pay the invoice, arguing that the CSM towed the lunar module during earlier missions.

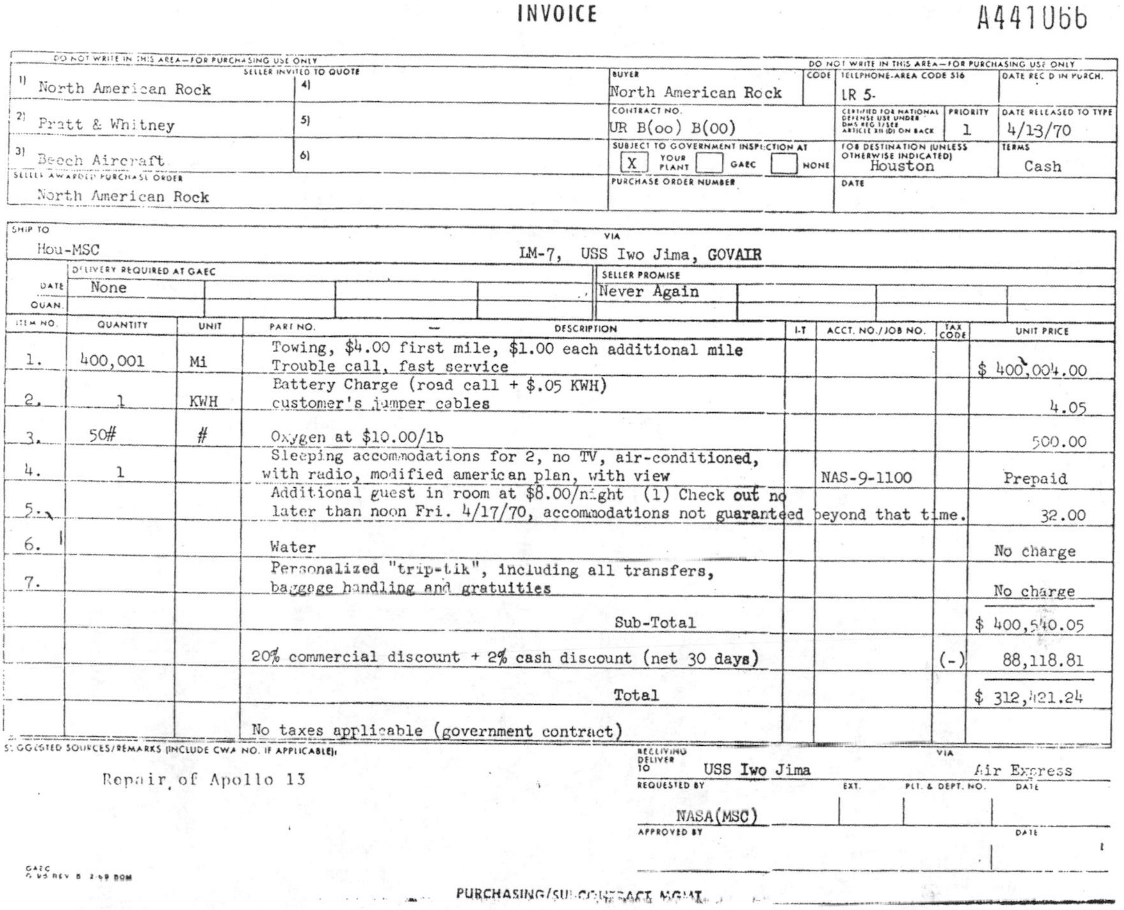

INVOICE   A441U66

DO NOT WRITE IN THIS AREA—FOR PURCHASING USE ONLY			DO NOT WRITE IN THIS AREA—FOR PURCHASING USE ONLY	

1) North American Rock   4)

2) Pratt & Whitney   5)

3) Beech Aircraft   6)

SELLER AWARDED PURCHASE ORDER
   North American Rock

BUYER
North American Rock

CONTRACT NO.
UR B(oo) B(00)

SUBJECT TO GOVERNMENT INSPECTION AT
[X] YOUR PLANT [ ] GAEC [ ] NONE

PURCHASE ORDER NUMBER

CODE | TELEPHONE-AREA CODE 516
LR 5-

CERTIFIED FOR NATIONAL DEFENSE USE UNDER DMS REG 1/1EE ARTICLE XII ON BACK | PRIORITY 1 | DATE RELEASED TO TYPE 4/13/70

FOB DESTINATION (UNLESS OTHERWISE INDICATED)
Houston | TERMS Cash

DATE

SHIP TO
Hou–MSC

VIA
LM-7,   USS Iwo Jima, GOVAIR

DELIVERY REQUIRED AT GAEC
DATE None
QUAN.

SELLER PROMISE
Never Again

ITEM NO.	QUANTITY	UNIT	PART NO. — DESCRIPTION	I-T	ACCT. NO./JOB NO.	TAX CODE	UNIT PRICE
1.	400,001	Mi	Towing, $4.00 first mile, $1.00 each additional mile. Trouble call, fast service				$ 400,004.00
2.	1	KWH	Battery Charge (road call + $.05 KWH) customer's jumper cables				4.05
3.	50#	#	Oxygen at $10.00/lb				500.00
4.	1		Sleeping accommodations for 2, no TV, air-conditioned, with radio, modified american plan, with view		NAS-9-1100		Prepaid
5.			Additional guest in room at $8.00/night (1) Check out no later than noon Fri. 4/17/70, accommodations not guaranteed beyond that time.				32.00
6.			Water				No charge
7.			Personalized "trip-tik", including all transfers, baggage handling and gratuities				No charge
			Sub-Total				$ 400,540.05
			20% commercial discount + 2% cash discount (net 30 days)			(-)	88,118.81
			Total				$ 312,421.24
			No taxes applicable (government contract)				

SUGGESTED SOURCES/REMARKS (INCLUDE CWA NO. IF APPLICABLE)

   Repair of Apollo 13

RECEIVING DELIVER TO   USS Iwo Jima   VIA   Air Express
REQUESTED BY   EXT.   PLT. & DEPT. NO.   DATE
       NASA(MSC)
APPROVED BY   DATE

GAEC
G VS REV 8 2-69 DOM

PURCHASING/SUBCONTRACT MGMT.

Figure 7.11 Towing bill sent by Grumman (constructor of the lunar module) to North American (command and service module contractor) after the safe return of Apollo 13. (Image courtesy of NASA.)

7.10 APOLLO 14 (JANUARY 31–FEBRUARY 9, 1971)

Call Signs: CM: *Kitty Hawk*; LM: *Antares*

Time on surface: 1 day 9 hours

The Apollo 14 crew, Alan Shepard (MC, Mercury 1), Stu Roosa (CMP) and Ed Mitchell (LMP) were sent to *Fra Mauro*, a hilly region on the edge of *Mare Imbrium*, and the intended landing site for Apollo 13. Shepard was the only member of the original Mercury roster of astronauts to go to the Moon.

While the launch was uneventful (if that is possible on top of a Saturn V), the crew had difficulty docking and extracting the lunar module as described in Chapter 5. Then, when in orbit around the Moon, and prior to the descent, *Antares* suffered problems with the onboard abort system necessitating a re-programming of the computer (Intermission 3).

When Apollos 11 and 12 landed, the detached lunar modules burned their descent engines to move into the lower descent orbit. The post-flight analysis of the Apollo 11 landing noted that more fuel than anticipated was consumed in the process of finding a safe landing spot. Hence ways of saving LM fuel were considered, to provide the crew with a safer fuel margin. Using the SPS to move the entire CSM–LM stack into the descent orbit before separation was one suggestion. However, given the lead-time needed to plan the burn and adjust the crew's mission timeline, incorporating that idea into Apollo 12 was deemed impossible. It was, however, put in place for Apollo 13 onwards. Apollo 14 was the first mission to carry out the manoeuvre.

During the powered descent, the LM landing radar failed to lock on to the surface, preventing the computer from receiving important data about the height and rate of descent (the fault was later traced to an unrelated bug in the radar's operation). Fortunately, cycling the radar's breaker (powering it on and off) allowed the system to acquire the reflected signal at 5.5 km altitude and Shepard was then able to land closer to target than any earlier Apollo mission.

During the two EVAs, the astronauts performed science experiments and collected samples (and hit an occasional golf ball). The region was thought to consist of ejecta from the impact that formed *Mare Imbrium* and contained relatively low ridges and hills. The crew also sampled ejecta material from *Cone Crater* formed by an impact thought to have unearthed[10] material from 80 m down.

Many mission samples turned out to be breccia, the geological term for rocks consisting of fragments from more ancient rocks – something that can happen when the heat and pressure of a meteorite impact fuse fragments together. One such sample named *Big Bertha* (Figure 7.12) is now thought to be of terrestrial origin based on the granite and quartz (common on Earth and very rare on the Moon) in its structure.

Mineral analysis suggests that the rock is roughly 4 billion years old. Its zircon content is far different from that found in other Moon rocks but very similar to the zircon found on Earth. Big Bertha is the oldest known terrestrial rock sample found to date. Presumably, Earth surface impact ejected fragments at escape velocity and Big Bertha was captured by lunar gravity. While this is a remarkable discovery, it is not unprecedented as 224 of the 61,000 meteorites found on Earth are most likely of Martian origin.

Figure 7.12 The Big Bertha sample prior to collection by the Apollo 14 crew. The rock is at the centre of the image casting a shadow to the left, between the wheel tracks of a small portable workbench that the astronauts used on the surface. (Image courtesy of NASA.)

7.11 APOLLO 15 (JULY 26—AUGUST 7, 1971)

Call Signs: CM: *Endeavour*, LM: *Falcon*

Time on surface: 66 hours 55 mins

Dave Scott (MC, Gemini 8, Apollo 9), Al Worden (CMP) and James Irwin (LMP), flew the first of the extended J missions (previous landings had H designations) that included revisions to the service and lunar modules. The heavier LM had the capacity to support a 75-hour surface stay and carried the lunar roving vehicle to the surface. A package of experiments added to the service module supported gathering lunar data from orbit. Worden had to perform the first deep space EVA on the return leg to retrieve a film canister from this science bay. The service module also carried a small sub-satellite that was released into lunar orbit shortly before departure. It was designed to study the plasma, particles and magnetic fields near the Moon. It functioned until January 1973, after which it probably crashed onto the lunar surface.

The heavier LM required a larger descent engine bell. The quantities of dust thrown off the Moon surface prevented Scott from seeing the surface as he moved below 18 m altitude. The astronauts had been drilled on the importance of cutting off the engine at the moment the contact light lit to indicate that the probes on the ends of the legs had touched the surface. Mission planners were nervous about exhaust gases reflecting back from the surface into the engine bell, with potentially explosive consequences. When Scott cut the power in response to Irwin's call of 'contact light', the LM was descending at 0.15 m/s and subsequently fell the remaining 0.49 m, impacting the surface at 2.1 m/s – the hardest of all the landings. Due to the obscuring dust, Scott inadvertently landed on the edge of a small crater which caused *Falcon* to tip in two directions by 7 to 8 degrees (Figure 7.13).

Figure 7.13 The Apollo 15 lunar module on the surface. Note the boot prints and rover tracks in the foreground, and the tilt of the module which caused some problems when extracting the lunar rover. (Image courtesy of NASA.)

After the landing, Scott opened the top hatch and stood on the ascent engine inside the lunar module, so that he could photograph the surroundings. This 'stand-up EVA' had been Scott's idea and was done only during Apollo 15. Scott was impressed by the geologists who trained the astronauts and heeded their advice to seek high ground and photograph the area where samples were to be taken. He reasoned that the lunar module would act as high ground. NASA required some convincing to allow the standing EVA due to the extra oxygen that would be vented from the LM. Scott won that concession and also persuaded NASA to let him carry extra camera lenses.

Deploying the lunar rover allowed the crew to venture further from the LM and it was used during three EVAs over a total of 27.9 km around the *Hadley Rill* region of the Moon (Figure 7.14). However, the astronauts were prohibited from venturing further than they could walk back, given the amount of oxygen carried in their PLSS backpacks.

EVA1 took them 67 m above the landing site where much of the surface lunar material was debris that slid from upper slopes. Most of the material collected consisted of regolith. EVA2 climbed higher along steeper slopes, posing traction problems for the rover. The astronauts found that they could not lean back far enough to get the top of the mountain in frame for pictures. They collected a crystalline rock fragment containing evidence of 4 billion-year old geology. The astronauts believed that they had seen evidence of a 'high water mark' where the original lava filled the mare plain.

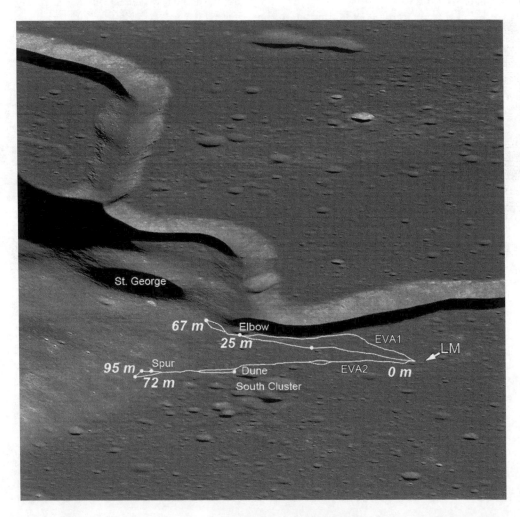

Figure 7.14 An image from the Lunar Reconnaissance Orbiter showing the Apollo 15 landing site and the traces of two EVAs. The numbers refer to elevations above the landing site. Hadley Rille is clearly visible. (Image courtesy of NASA, Goddard Space Flight Center and Arizona State University.)

The third EVA was spent obtaining core samples by drilling into the surface. The sampling started the previous day but the drill got stuck in the surface. The crew then took the rover to the rille to sample bedrock. At the end of this EVA, Scott performed the hammer and feather demonstration described in Chapter 2. The lunar rover was parked a distance away from the LM and was successfully used to remotely televise the ascent stage liftoff.

The crew gathered valuable scientific data and demonstrated the effectiveness of the enhanced lunar module to extend time on the surface. Longer surface stays, along with the use of the lunar rover, pushed exploration into a higher gear on subsequent missions.

The splashdown completed safely even though one of the main parachutes did not deploy properly. The Apollo 15 crew were the first crew not be quarantined on their return.

7.12 APOLLO 16 (APRIL 16–27, 1972)

Call Signs: CM: *Casper*, LM: *Orion*

Time on surface: 2 days, 23 hours

The CMP for Apollo 16, Ken Mattingly, had been assigned to Apollo 13 until he was 'bumped' in favour of Jack Swigert, over concerns that Mattingly might develop the measles in flight after exposure via another member of the astronaut team. While Mattingly never developed measles, he did work closely with John Aaron ('SCE to auxiliary') in developing the vital power-up sequence for the Apollo 13 command module to enable it to survive on battery power through re-entry.

The Apollo 16 landing site was selected to allow John Young (MC, Gemini 3, Gemini 10, Apollo 10) and Charley Duke (LMP, CAPCOM for the Apollo 11 landing) to sample material older than that obtained in three of the first four landings. Apollos 11, 12 and 15 touched down on or near maria that were filled with younger material. Apollo 14 explored a hilly region, but Apollo 16 was the first mission to visit genuine highlands. With no new equipment or procedures to test, surface time could be devoted to scientific operations.

The central highland regions of the Moon were originally thought to have been formed by a volcanic process similar to those on Earth. Scientists hoped an expedition to the area would confirm that theory. The selected site in the *Descartes* region was between two young impact craters, *North Ray* and *South Ray*.

These visibly lighter craters were targeted because they contained exposed bedrock that could be sampled. Their lighter colour suggested that the ejecta material had not yet been darkened by prolonged exposure to solar radiation and micrometeorites, indicating that it was comparatively young. The astronauts sampled material from a ray near *South Ray* crater during EVA2 and visited *North Ray* during EVA3 (Figure 7.15). A substantial scientific achievement of the mission was showing that the highland region was formed of impact breccia and was not volcanic.

Earlier in the mission, during preparation for the Moon landing, a malfunction occurred in the SPS backup systems. The lunar module had only just separated from the CSM and mission rules suggested that the craft should link up again in case the landing had to be aborted and the descent engine of the LM used to break lunar orbit. Mission control decided after hours of analysis that the error had a satisfactory work-around and the mission continued. However, the burn to achieve the correct orbit for the particles and fields sub-satellite was cancelled. As a result, its orbit became unstable and it crashed 35 days after deployment. The SPS burn to break orbit was carried out without a hitch.

7.13 APOLLO 17 (APRIL 16–27, 1972)

Call Signs: CM: *America*, LM: *Challenger*

Time on surface: 3 days 3 hours

Apollo 17 represented the first night-time and final manned launch of the Saturn V. The rocket did fly again, in an unmanned capacity, to deliver the *Skylab* Space Station to orbit.

As the final lunar exploration mission (a record that regrettably stands to this day), pressure was exerted on NASA by the scientific community to ensure that a professional geologist was on the flight. The lunar module pilot, Harrison Schmitt, is a PhD geologist from Harvard University and worked for the US Geological Survey's Astrogeology Center on developing field techniques for the Apollo crews before he

Figure 7.15 Young on the slope of *North Ray* crater during EVA3. (Image courtesy of NASA and Charles M. Duke Jr.)

joined NASA in 1965. Schmitt (LMP) was assigned to Apollo 17 along with Eugene Cernan (MC, Apollo 10) and Ronald Evans (CMP).

The *Taurus Littrow Valley* landing site was selected so that Schmitt and Cernan could acquire old highland material, specifically from the remains of a landslide from the south wall of the valley (Figure 7.16) and material from a young volcanic event. As with Apollo 15, the site was on the edge of a mare, but this drawback was outweighed by other advantages of the area.

During the outward journey, the crew repeated and extended an experiment first performed by Apollo 16. On all previous missions, crew members reported seeing 'streaks' or 'specks' of light when the cabin lights were low and (strangely) even when their eyes were closed. One of the crew wore a device that recorded the times, strengths and paths of any high energy particles that passed through it. The data could then be correlated with any reported light flashes, establishing that electrically charged particles passing through the retina were most likely the cause.

The third and last EVA of the Apollo era (Figure 7.17) started at 09:26 UTC on December 13, 1972.

Cernan and Schmitt drove the lunar rover to explore the base of the *North Massif*, the *Sculptured Hills* and the *Van Serq* craters. After unveiling a plaque on the LM, Gene Cernan became the last human (so far) to leave the surface of the Moon and he commemorated the event with the following thoughts:

> …I'm on the surface; and, as I take man's last step from the surface, back home for some time to come—but we believe not too long into the future—I'd like to just [say] what I believe history will record. That America's challenge of today has forged man's destiny of tomorrow. And, as we leave the Moon at Taurus-Littrow, we leave as we came and, God willing, as we shall return, with peace and hope for all mankind. Godspeed the crew of Apollo 17.

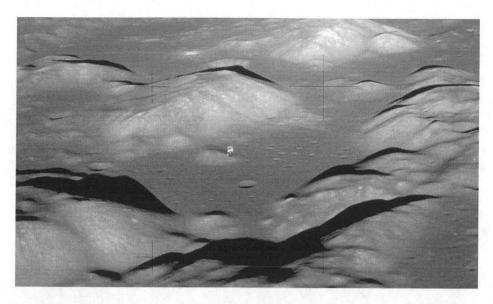

Figure 7.16 A view of Apollo 17's Taurus-Littrow Valley landing site from the lunar module *Challenger*. Note the pale landslide material (bottom left quadrant of the top right fiducial cross) and the CSM combination (centre). (Image courtesy of NASA.)

Figure 7.17 Lunar geologist Harrison Schmitt next to a large split boulder during Apollo 17's EVA 3. The lunar module can be seen in the distance. (Image courtesy of NASA and Eugene Cernan, annotation by the author.)

Figure 7.18 Locations of six landing sites of Apollo missions. (Image courtesy of Wikimedia Commons, distributed under Creative Commons Attribution Share Alike 3.0 Unported. https://creativecommons.org/licenses/by-sa/3.0/.)

Splashdown took place on December 19, 1972 at 19:24:59 UTC, bringing the Apollo Moon missions to a close (Figure 7.18).

7.14 APOLLO-SOYUZ (APOLLO: JULY 15–24, 1975; SOYUZ: JULY 15–24, 1975)

By 1975, the cooling of the Cold War and the policy of Détente being pursued at the time created the right climate for a joint Soviet–American mission that involved docking an Apollo command and service module with a Soyuz capsule. The Apollo crew members were Tom Stafford (MC, Apollo 10), Vance Brand (CMP) and Deke Slayton (Docking Module Pilot). Having been selected for the Mercury programme in 1959, Slayton was never permitted to fly because of an irregular heart rhythm that came to light via electrocardiogram tests during training. He became head of the astronaut office, a role that included responsibility for crew assignments. Over the years, he took several steps to improve his condition and after examination at the Mayo Clinic in 1970 he was restored to flight status.

The Soviet capsule was commanded by Alexei Leonov, the first human to walk in space. He was accompanied by Valeri Kubasov who had flown a previous Soyuz mission. The crews spent some of their time in space conducting experiments including using the Apollo to create an artificial solar eclipse that allowed photography of the Sun's corona from the Soyuz.

ENDNOTES

1. The main report comprised a 20 cm thick, 3,000 page stack of 14 booklets.
2. This complex system was designed to minimise the risk of hatch failure in the vacuum of space. Redundancy was assured by installing multiple hatches. By using an inner hatch that opened inward, the designers could be sure that the pressure inside the capsule would force the hatch shut. The astronauts were happier with a simpler arrangement that allowed them to exit quickly.
3. A complete and harrowing account of the events of that day can be read in *Moon Shot* by Alan Shepard and Deke Slayton. Open Road Media, Revised ed., May 2011.
4. Not to say lives, of course.
5. CAPCOM is the acronym for Capsule Communicator – an astronaut responsible for the direct radio communications between Earth and a spacecraft. Only the CAPCOM was authorized to relay instructions.
6. The story of this project can be found at https://www.honeysucklecreek.net/Apollo_11/tapes/Apollo_11_Tape_Search_Flyer.pdf and extracts from the completed restoration can found at https://www.nasa.gov/multimedia/hd/apollo11_hdpage.html
7. M Collins, *Carrying the Fire*, Anniversary Edition, Farrar, Straus & Giroux, July 2009.
8. For his quick thinking and cool head, Aaron was given the unofficial NASA title 'steely-eyed missile man'. Aaron went on to play an important role in saving the Apollo 13 crew. The phrase 'steely-eyed missile man' appears from time to time in popular culture as a nod to Aaron's intervention.
9. J Lovell and J Kluger, *Apollo 13*, Houghton Mifflin, April 2002.
10. Unmooned?

8

The space shuttle

8.1 ROCKET PLANES

In the late 1950s, the United States Air Force tested rocket planes (such as the X-15, Figure 8.1) that could fly higher than 80 km (the NASA criterion for the pilot being classed as an astronaut). It is possible that rocket planes could have been refined into a means of getting into orbit, but NASA took the view that their development would not be completed soon enough to enable Project Mercury to beat the Russians into space.

With the race to the Moon won, NASA was able to reconsider a system along the lines of a space-plane that could shuttle back and forth into space. Potentially, a vehicle of this type would allow a number of elements to be reused for flight after flight eventually bringing launch costs down, in turn enabling the commercial exploitation of space. During the Nixon administration, NASA started to divert funds from Apollo to the development of what became the space shuttle.

The *Space Transportation System* (STS), or *Space Shuttle* for short was a flawed, beautiful, complex and innovative machine (Figure 8.2). It combined the aerodynamic characteristics of an aeroplane that can glide, land and be reused, with rocket technology to provide the thrust required to get into orbit. However, as the shuttle was designed in the 1970s and first flew in 1981, it was fundamentally limited by the technology available at the time.

Figure 8.1 The North American X-15 rocket plane which flew from 1959 to 1968. Neil Armstrong was one of the test pilots for this aircraft. (Image courtesy of NASA.)

Figure 8.2 The first shuttle on its mobile launch pad. Note that the external tanks on the early flights were painted white to match the orbiter. The large structure to the left is the payload changeout room that swung around to mate with the orbiter and provide a weather-tight seal so that the payload bay doors could be opened. The vertical gantry at the far left is the support for the structure. At the bottom of the gantry is a cab for the driver when the structure was driven around on a track in an arc to mate with the orbiter. (Image courtesy of NASA and Kennedy Space Center.)

8.2 THE STS COMPONENTS

There were three components to the space transportation system. The *orbiter* was the plane-like component that rode into orbit on the back of a large orange-coloured cylinder (the *external tank*). Inside the external tank were two vessels, one containing liquid oxygen (LOX) and the other liquid helium (LH_2). Both liquids were pumped to the main shuttle engines situated at the back of the orbiter.

Strapped to the sides of the external tank were the *solid rocket boosters* (SRBs). These long cylinders contained moulded solid propellant that generated 71% of the thrust at launch. Without the SRBs, the shuttle would never have left the ground. The boosters were jettisoned after 2 minutes of flight. By then, the mass had reduced sufficiently (due to consumption of propellant in the external tank) for the main engines to continue accelerating the craft into orbit by themselves. The SRBs parachuted into the ocean, to be recovered and used again. Each SRB casing was designed to last 20 launches. The orbiters were intended for a minimum of 100 launches. The empty external tank was jettisoned to burn up in the atmosphere (Figure 8.3). It was the only part of the shuttle system that was not re-used.[1]

The orbiter could carry a crew of up to seven people[2] and remain in space for several days (the record was 17 days and 15 hours for STS-80). It could deploy satellites into orbit, capture and return faulty satellites to Earth, transport scientific experiments, repair satellites such as the *Hubble Space Telescope in situ* and eventually transport modules for the *International Space Station* (ISS).

On return to Earth, the orbiter oriented itself so that its underside faced down and slightly forward. It then used its supplementary engines (the *Orbital Manoeuvring System* engines that used hypergolic propellants and were sited either side of the main engine cluster) to slow down and trigger a gravity-assisted descent into the atmosphere. The path followed by the orbiter was complex and designed to minimise the effects of air resistance and thermal excess on the craft. With the shuttle's nose pointing ~40 degrees up compared to the

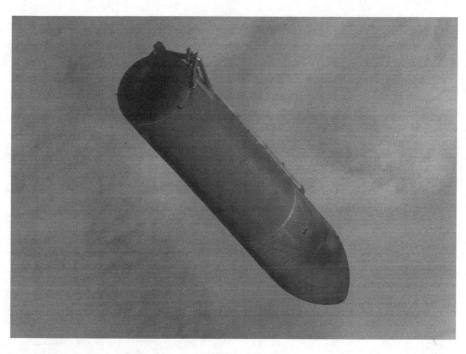

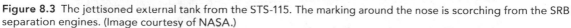

Figure 8.3 The jettisoned external tank from the STS-115. The marking around the nose is scorching from the SRB separation engines. (Image courtesy of NASA.)

direction of flight, the wings generated significant lift as the orbiter entered the denser atmosphere at high speed. As this would tend to lift the vehicle out of the atmosphere, the flight plan called for left and right rolls to direct the lift more horizontally than vertically which allowed the rate of descent to be regulated.

Most of the descent path was flown automatically by the onboard computer systems, but the pilots were able to take over in case of an emergency. After passing through the re-entry phase, the shuttle glided at supersonic speeds to its landing strip as it did not possess engines that could guide it through the atmosphere. The orbiter touched down on a runway like a conventional aeroplane (but somewhat faster – 341 to 364 km per hour!) and came to a halt with the aid of a parachute.

8.2.1 THE ORBITER

Over the period of the space shuttle's service, six orbiters were constructed including the original glide-test craft named *Enterprise* and *Endeavor,* which was built after the first shuttle accident destroyed *Challenger* (Table 8.1).

The orbiter consisted of several major sections as shown in Figure 8.4. Starting from the front of the craft, the *forward fuselage* (extending from the nose back to the payload bay doors) had upper and lower sections that fit together like a clamshell. The forward RCS module contained thrusters for manoeuvring in space. Additional thrusters were mounted in the blisters on both sides of the tail fin (vertical stabiliser).

The pressurised *crew compartment* was a welded aluminium structure with four attachment points that held it inside the forward fuselage. The small number of attachments minimized heat loss in space and the thermal conduction from the craft's exterior to the compartment during re-entry.

Access to the crew compartment was via the main entry hatch on the left side of the orbiter (as viewed from above and shown in Figure 8.7). On the mobile launcher, the white room at the end of the swing arm connected with this hatch, allowing crew access prior to launch. In an emergency, the hatch could be blown open. The second way in and out of the crew compartment was via an airlock that passed through the pressure vessel and the rear bulkhead of the forward fuselage, giving access to the huge payload bay. On some shuttle missions, an additional pressurised module known as *Spacelab*, *Spacehab* or a docking adapter was mounted in the payload bay. The crew could float into any module via a tunnel linked to the airlock.

The crew compartment was divided into two decks (Figure 8.5). The top level (effectively the blister on the upper surface of the orbiter) from the windows back to the payload bay doors was the flight deck where the

Table 8.1 Summary of Orbiter flights from 1977 through 1992

DESIGNATION	FIRST FLIGHT	LAST FLIGHT	NUMBER OF MISSIONS	DESCRIPTION
Enterprise	August 1977	October 1977		Used for atmospheric test flights; launched from a modified Boeing 747; on display at Intrepid Sea Air & Space Museum, New York
Columbia	STS-1 1981	STS-107 2003	28	First space-worthy orbiter delivered; destroyed during STS-107 re-entry
Challenger	STS-6 1986	STS-51-L 1986	10	Destroyed during launch of STS-51-L
Discovery	STS-41-D 1984	STS-133 2011	39	Currently on display at Steven F Udvar-Hazy Center, Virginia
Atlantis	STS-51-J 1985	STS-135 2011	33	Flew last space shuttle mission; currently on display at Kennedy Space Center Visitor Complex
Endeavour	STS-49 1992	STS-134 2011	25	Built to replace *Challenger* and constructed from spare parts built for *Discovery* and *Atlantis*. Currently on display at California Science Center, Los Angeles

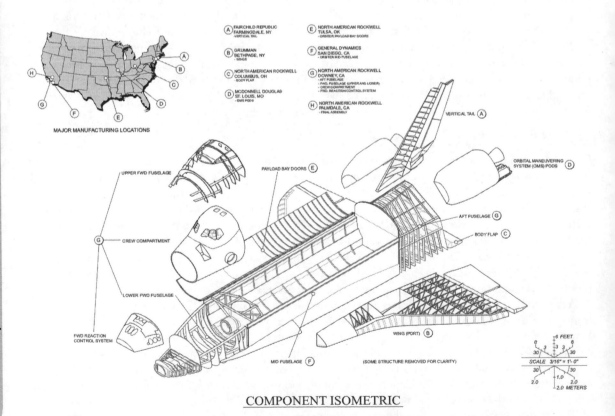

COMPONENT ISOMETRIC

Figure 8.4 The main components of the orbiter. (Image courtesy of NASA.)

The space shuttle

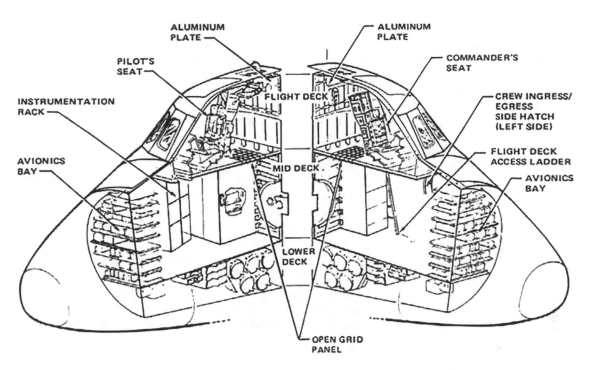

Figure 8.5 A cross-section through orbiter's crew compartment showing the decks. (Image courtesy of NASA.)

commander and shuttle pilot sat behind the control panel during launch. Behind their seats was a work area with two viewing windows looking back along the spine of the orbiter and another pair that faced vertically upward. The deployment of satellites, payloads and the remote manipulator arm could be controlled from the work area, which also contained controls for manoeuvring the shuttle. In some configurations, two additional seats were mounted on the flight deck during launch and later stowed away. An opening in the floor of the top deck led to a ladder extending down to the mid deck; of course the ladder was superfluous in zero-*g*.

The mid deck could house three or four more seats during launch, depending on which orbiter was in use, and served as the main living area for the crew. It also contained three equipment bays. The two at the front of the deck occupied the entire forward bulkhead wall and the one at the back was to the side of the airlock. The lockers in these bays could hold the crew's personal items as well as mission equipment. A sophisticated 'waste management system' that could cater for both male and female clients (unlike that on Apollo) was next to the main hatch.

Depending on the space required for equipment, a galley and a washing unit could be installed on the mid deck. If a galley was not possible, a suitcase-sized food warmer was installed. Ice cream and frozen steaks were sometimes included in the menus if a refrigerator used for biomedical experiments was available (it was important to mark all contents clearly!). The crew could sleep in their seats, in sleeping bags tethered to the wall or, if there was room, in bunk beds mounted on the mid deck.

Inside the crew compartment, the environmental systems provided a nearly normal mixture of 80% nitrogen and 20% oxygen at sea level pressure. The air was filtered and recycled. During launch, the crew wore pressure suits very similar to those worn by fighter pilots. While on an EVA, astronauts wore shuttle spacesuits evolved from the earlier Apollo design. For day-to-day mission operations, the crew wore jeans and T-shirts or other personal clothing.

The *mid fuselage* was 18 m long and was the main load carrying portion of the orbiter. It transmitted the force of the engines from the rear fuselage to the rest of the craft and contained the structure that linked the wings together. At the top of the mid fuselage sat the *payload bay* with its large hinged doors constructed from a graphite epoxy frame with honeycomb panels. The doors were opened when the craft was in orbit, regardless of whether the bay contained a payload, as their underside carried the thermal radiators for the orbiter. For most of the mission, the orbiter circled the Earth with the payload bay open and facing towards the ground. This ensured that much of the direct sunlight fell on the underside where the most heat-resistant surfaces were placed.

After the mid fuselage came the *rear fuselage* where the main shuttle engines that burned LOX and LH$_2$ from the external tank and two smaller orbital manoeuvring system (OMS) engines were mounted. The vertical tail fin with two flaps that opened outward to work as air brakes on landing was also attached to this section.

The wings were constructed of an aluminium alloy with a honeycomb cover.

8.2.1.1 The remote manipulator arm

The first remote manipulator arm flew on the *Columbia* orbiter in 1981 and was a gift from Canada to NASA. Additional arms ordered by NASA had a total export value of $400 million. Primarily, the arm was designed to grab satellites and guide them into the payload bay. The arm could also deploy objects from the bay. During the four missions[3] to repair and service the *Hubble Space Telescope* (Figure 8.6), the arm linked to the telescope, pulled it into the payload bay, carried the astronauts (functioning as a mobile work platform) and finally delivered the refurbished telescope back into orbit. Other more unconventional uses included knocking away a block of ice from a clogged waste-water vent that might have endangered the shuttle upon re-entry, pushing a faulty antenna into place, and successfully activating a satellite (using a swatter made from briefing covers) that failed to reach the proper orbit. The arm was guided from the station at the back of the flight deck, using translational and rotational hand controllers.

Although the arm could not lift its own weight on Earth, it accurately manoeuvred payloads of 30,000 kg in orbit.

Figure 8.6 The *Hubble Space Telescope* (large cylinder at the centre) was repaired and refurbished by the crew of STS-61 in December 1993. In this image, one crew member is being moved on the end of the shuttle's remote manipulator arm while another works in the shuttle payload bay. (Image courtesy of NASA.)

The space shuttle

8.2.1.2 Thermal protection

Before the space shuttle orbiter went into service, spacecraft survived re-entry by using ablative heat shields. As the orbiters were intended to be re-used up to 100 times, a new way of protecting them from the heat was required.

The orbiter's nose, the area under its chin and the leading edges of its wings encountered the most friction during re-entry. About 20 minutes before touchdown, their temperature climbed as high as 1,600°C. These areas were protected by a carbon composite, moulded in a single piece for the nose and in several sections for the leading edges of the wings. Some areas were further insulated with quartz blankets and metal foils. The composite was constructed from a graphite cloth soaked in a special resin. Layers of the cloth were then laminated together and the whole structure baked in an oven that converted the resin into carbon. A mixture of aluminium, silicon and silicon carbide was then coated over the top. Much of the remaining surface of the orbiter was insulated by a combination of 24,300 tiles and 2,300 flexible insulation blankets (Figure 8.7).

There were two types of tile used: (1) black high-temperature reusable surface insulation (HRSI) tiles covered the underside of the orbiter, the region around the flight deck windows and various other areas and (2) white low-temperature reusable surface insulation (LRSI) tiles were originally used extensively on the areas of the orbiter that did not have to survive extreme heating. Subsequently, most of the LRSI tiles were replaced by flexible insulation blankets (FIBs) made from a waterproofed quilted fabric consisting of silica felt between two layers of glass cloth sewn together with silica thread. The blankets were developed after *Columbia* was delivered to NASA and first used around the OMS pods on *Challenger*. They were used more extensively from *Discovery* on and replaced many of the LRSI tiles on *Columbia* after the *Challenger* accident.

The FIBs had better durability and cost less to make and install than the tiles. They were used on the upper sides of the orbiter fuselage, sections of the payload bay doors, most of the tail fin and parts of the upper wings and around the observation windows. After each flight, it was necessary to replace the waterproofing agent that burnt away during re-entry. This was done by injecting new agent into each tile through existing small holes.

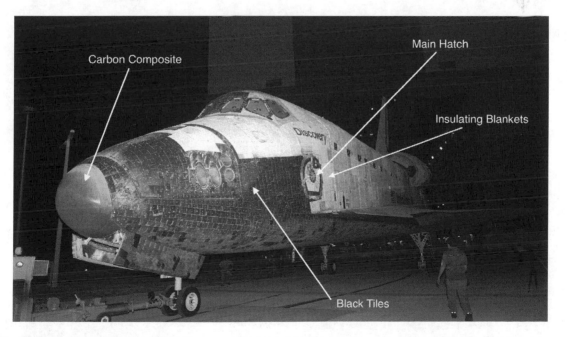

Figure 8.7 This close-up of *Discovery*'s nose shows some of the thermal protection materials used. (Image courtesy of NASA with the author's additions.)

The tiles were made from fibres of pure white silica refined from sand. These fibres were mixed with high purity de-ionised water and other chemicals, then poured into a mould. The excess liquid was squeezed out of the mixture and the blocks were dried in a microwave oven at the Sunnyvale plant of Lockheed Space Operations Co. They were then sintered in a 1,290°C oven that fused the fibres without melting them.

Each tile on the orbiter had a curvature on its underside that exactly matched the curve on the orbiter's body at the point where the tile was applied. As a result, each tile was a unique shape and could only be placed in its specified location. The blocks were first cut with saws and then milled to the correct shape by diamond-tipped machines. Full documentation existed for each tile, and the manufacturing process could be traced back to the individual batch of sand used.

The final step was to coat the tiles with borosilicate glass into which various chemicals were added to provide colour and heat rejection properties. Surface heat dissipated so quickly that a tile could be held by its corners a few seconds after removal from the sintering oven while its centre still glowed red. The tiles were delicate and each one had to be inspected, repaired or replaced after every mission.

8.2.2 THE EXTERNAL TANK

The external tank was the largest space shuttle component, being 47 m long with a maximum diameter of 8.5 m, and once loaded with propellant it became the heaviest. Internally, the cylinder was divided into a LOX tank at the top, an inter-tank structure in the middle and an LH_2 tank at the bottom.

The external tank was connected to the orbiter at three mounting points. The forward attachment was under the shuttle's nose, about halfway between the tip and the flight deck windows. The two aft attachments were on either side of the tank and connected to the orbiter on the mid fuselage, just behind the ends of the wings. In the same region, umbilical lines carried propellant, gases, electrical signals and electrical power to and from the orbiter. The control signals for the solid rocket boosters also passed through these umbilicals.

8.2.3 THE SOLID ROCKET BOOSTERS (SRBs)

The SRBs split into several segments as shown in Figure 3.1. The top three segments contained the control electronics, parachutes and the igniter that sent a 45 m jet of flame down the length of the tube to set fire to the propellant on liftoff.

Next were the four segments packed with the solid propellant mixture. By mass, the mixture proportions were 69.6% ammonium perchlorate oxidiser, 16% aluminium fuel, 0.4% iron oxide catalyst, 12.04% polymer binder to hold the mixture together and 1.96% epoxy curing agent.

The contractor moulded the propellant into comparatively short segments and then stacked them into the booster frame. If they had produced one long column of mixture instead, the heavier components would have sunk to the bottom as the propellant took more than 5 hours to cure. Matched pairs of boosters were produced for each flight by making sure that two SRBs were loaded with propellant from the same batch. This was crucial: if one booster had more thrust than the other, the craft might not have left the ground vertically!

The first propellant segment was moulded into an 11-point star with the points facing inward to give a large surface area. The remaining segments were internally shaped as a truncated cone. This allowed the boosters to provide maximum thrust at ignition followed by a decrease (down by about one-third) 50 s after liftoff, which corresponded to the time when the shuttle would start to feel maximum pressure from air resistance. The lowest segment also contained the convergent–divergent nozzle that could gimbal up to 7 degrees to steer the shuttle during ascent. The next component was a skirt assembly that handled the weight of the entire shuttle when it rested on the mobile launcher. The skirt also contained four small motors designed to separate the boosters from the shuttle about 2 minutes into the flight. Four more such motors sat at the top of the booster stack.

Great care had to be taken in joining the segments together as the weight of the packed fuel mixture could distort the casing, making it hard to align the parts. Most of the joining was done at the factory, but the *field joints* were made and sealed inside the vehicle assembly building (VAB). A completed SRB was over 45 m tall, 371 cm in diameter and weighed 590 tonnes at launch.

The space shuttle

8.3 COMPUTER SYSTEMS

The shuttle orbiter had fully digital fly-by-wire avionics and was one of the first flying machines to employ such a system.

Adding to the complexity of the computer systems was the switch to more autonomous control of certain aspects of the ship's performance formerly governed by ground-based systems, such as fuel levels and life support. Finally, as the crew had to interact with the computers far more frequently than was the case on previous spacecraft, significant amounts of software and hardware were devoted to displaying data and commanding the computers.

The five computers onboard the orbiter operated redundantly. Off-the-shelf designs from IBM were chosen, rather than machines built for specific applications, as had been the case with Gemini and Apollo. However, the input and output circuits and memory components were purpose made. Another change from Apollo was the switch to a high-level programming language.[4] Previous systems were programmed using machine level language that was expensive and difficult to debug. A specialist programming language known as HAL/S[5] was developed for the shuttle computers.

The earliest shuttle computers were IBM AP-101 central processing units that handled 480,000 instructions per sec. They shared access to the core memory (104 K on the first flight) and the re-writeable mass memory units.

8.4 ASSEMBLING THE SPACE SHUTTLE

The process of preparing a shuttle for flight began with the retrieval of the spent solid rocket boosters. Two dedicated ships (*Liberty Star* and *Freedom Star*) sailed to where the SRBs splashed down and towed them back to shore (Figure 8.8). The boosters were then inspected, cleaned and parts of them shipped back to the manufacturer in Utah to be refilled with solid propellant. The segments then returned to the Kennedy Space Center by rail (they were too dangerous to fly) and re-used in the assembly of another shuttle.

Figure 8.8 Towing a recovered SRB back to shore. (Image courtesy of NASA and Kennedy Space Center.)

First, the lowest segments of the SRBs were mounted on support posts atop the mobile launch platform (Figure 8.9). Then the remaining segments were stacked in sequence until two complete SRBs stood on the platform. Assembling the boosters was time-consuming and difficult. The connecting bolts and joints had to be rigid enough for the entire stack to function as a complete unit during the stress of launch.

Each external tank was only used once and the manufacturer delivered a new one for each launch. The tanks traveled from New Orleans to Cape Canaveral by barge, being too large to fit on a truck bed or inside an aeroplane. A gigantic crane inside the VAB lifted the external tank above the mobile launcher and lowered it between the two standing SRBs. The crane supported the mass of the empty tank while engineers bolted it in place. This operation required several days of painstaking work to ensure accurate alignment.

While all this was going on, the orbiter was in preparation at the *Orbiter Processing Facility* (OPF) built next to the VAB after the Apollo programme ended (Figure 8.10). The OPF consisted of three almost identical 31 m tall buildings containing hangars and maintenance workshops. Three orbiters could be serviced in the OPF at one time.

Figure 8.9 Moving a pair of stacked SRBs atop a mobile launcher from one high bay in the VAB to another. (Image courtesy of NASA and Ben Cooper.)

Figure 8.10 The space shuttle orbiter *Discovery* is backed out of Orbiter Processing Facility 3 as part of the 'rollover' to the nearby vehicle assembly building where it was mated with an external tank–SRB combination prior to launch. (Image courtesy of NASA and Dimitri Gerondidakis.)

The first shuttle flights ended with a landing on one of the extra-long runways at Edwards Air Force Base in California. The shuttle was then loaded onto a converted 747 jumbo jet and flown back to Kennedy Space Center in Florida. The first orbiter to land at Kennedy directly was *Challenger* at the end of the tenth mission (STS-41-B). Kennedy became the routine landing site from STS-40 onward. The jumbo jet was still available if bad weather prevented a shuttle from landing at Kennedy.

After a shuttle returned to Kennedy, it was wheeled into the OPF where an intensive inspection and service routine lasting 4 to 6 weeks began. The craft was jacked up off its landing wheels and surrounded by a complex system of scaffolding designed to allow access to every part of its body. Each heat-resistant tile was inspected by hand and damaged or missing tiles were replaced. The main engines and fuel pumps were removed and sent for refurbishment. Sometimes the same engines were re-installed; on occasion they were transferred from a different orbiter or new ones were fitted. Thorough checks were made of all electrical circuits, computer systems, guidance systems and life support components. The inside of the craft was also thoroughly cleaned!

After passing the inspection and refurbishment processes, the orbiter was lowered onto its landing gear and wheeled to the VAB. Meanwhile, the mated external tank and SRBs on the mobile launch platform were surrounded by another scaffolding array allowing the technicians to access as much as possible. As a result, the orbiter had to be lifted off its landing gear into a vertical position and then winched 100 m over the top of the scaffold and down the other side to mate with the other components. The crane held the orbiter while it was bolted to the external tank (Figure 8.11). After that, the entire shuttle assembly was supported by the solid rocket boosters.

The next stage was to drive the crawler vehicle into the VAB and under the mobile launcher. Jacks were used to lift the launcher clear of its supports and it drove out on the crawler as the Saturn Vs had in the Apollo era. The crawler and mobile launchers were modified versions of the Apollo components (Figure 8.12).

Figure 8.11 An orbiter supported by a crane during attachment to the external tank. The curved base of the tank can be seen behind the shuttle's main engines. The SRBs are visible on either side of the orbiter above the leading edges of the wings. (Image courtesy of NASA.)

8.5 INSTALLING THE PAYLOADS

The orbiter was capable of carrying a variety of payloads in its bay. Large structures that required direct connections to the shuttle for power or life support were loaded on board while the orbiter was horizontal in the OPF. Examples of such payloads were *Spacelab*, *Spacehab* and experiment pallets to be used in orbit. Other payloads, generally satellites to be deployed in orbit, were installed into the payload bay while the shuttle was at the launch pad. The *Payload Changeout Room* referred to in the caption for Figure 8.2 was instrumental in this operation. It was large enough for the orbiter's payload bay doors to be opened within the room.

First, the payload was loaded into a vertical canister and hoisted into the room while it was in its open position. Then the structure closed on the orbiter to provide a clean room in which the payload underwent final preparation before being placed on board the shuttle. The ability to install payloads on the pad provided extra flexibility in loading items onto any shuttle that was ready for launch.

8.6 LAUNCHING A SHUTTLE

Like the Saturn V, the space shuttle required many hours of preparation to ensure a safe launch. The countdown sequence generally started at T minus 48 hours – in actuality about 72 hours from launch; NASA factored in several pauses (holds) to allow for delays and faults. Some of the more interesting specific events in the countdown sequence were:

- The crew left their quarters at T minus 2 hours and 55 minutes. After taking a lift to the top of the service tower next to the shuttle, they crossed the gantry to the white (clean) room mated to the external hatch of the orbiter. A team of technicians helped them strap into their couches for the launch. As the orbiter was standing vertically, the crew were forced to lie on their backs in heavy pressure suits for the remainder of the countdown.

Figure 8.12 The completed shuttle stack with *Atlantis* rolls out of the VAB on its mobile launch pad. (Image courtesy of NASA.)

- Over the next couple of hours, the team in the launch control building next to the VAB monitored the weather, pressure in the propellant tanks and fuel lines, and numerous other items. Inside the orbiter, the commander and pilot checked the systems and aligned the guidance controls.
- A built-in hold took effect at T minus 20 minutes. This short delay allowed the launch director to give last-minute instructions to his team in the control room. The final pre-planned hold was scheduled at T minus 9 minutes when the launch director asked each system manager for final launch approval. Once the 'go' instruction was given, the countdown was under the control of the automatic launch sequencer and the pace of activity increased rapidly.
- At T minus 7 minutes and 30 seconds, the crew access gantry with the white room swung away.
- At T minus 6 minutes, the auxiliary power units on the orbiter powered up to provide power for engine gimballing.
- At T minus 4 minutes and 30 seconds, the orbiter switched to electricity generated by its fuel cells rather than power provided by ground-based systems.
- From T minus 4 minutes to T minus 3 minutes, the orbiter's main engines were automatically gimballed to ensure that they functioned properly. After the test, the engines moved to their launch positions.
- At T minus 1 minute and 57 seconds, the cap on top of the external tank, which allowed boiling LOX to vent away, was retracted.
- At T minus 31 seconds, the orbiter's onboard computers took over the countdown.
- At T minus 11 seconds, water was pumped into the trench below the launch pad and also to the sprinklers that sprayed water across the top of the launch platform. This was another modification of the Apollo system. The incredible sound pressures built up by the SRBs on launch could have reflected off the metal surface of the platform and damaged the orbiter. Free-flowing water across the surface

dampened the sound waves and minimised the problem. The first orbiter to launch lost 16 tiles and another 148 were damaged, mainly due to pressure waves generated at launch. That led to modification of the water suppression systems.

- At T minus 8 seconds, sparks appeared beneath the main engine nozzles. As the engines ignited within seconds, it is often assumed that the sparks were the trigger. In fact, the flow of LH_2 to the engines began some seconds before the LOX was released and created a hydrogen-rich environment in the thrust chamber. Some of this hydrogen tended to leak and could have formed a highly explosive pool under the engine bells. The sparks burned off the hydrogen just before main engine ignition.
- At T minus 6.6 seconds, the first main engine ignited.
- At T minus 6.48 seconds, the second main engine ignited.
- At T minus 6.36 seconds, the final main engine ignited. The three engine starts were staggered to gradually increase the load on the shuttle rather than force it to handle the full strain at once. The force of the main engines caused the top of the shuttle to tip by 1 to 2 m. The super-hot exhaust of the main engines entered the trench beneath the mobile launcher, flash heating the water there into the steam that billowed around the base of the pad in the moments before launch.
- As soon as all the main engines operated at 90% or more of their nominal thrust, the computers signalled the SRBs to ignite. At this point, a launch could not be halted: the shuttle was heading into the air come what may. The exhaust from the SRBs formed a bright jet of flame, unlike that from the main engines which, being super-hot water vapour, was invisible (as was the exhaust from the Saturn V second and third stages).
- As the shuttle climbed above the gantries and scaffolding of the mobile launcher ('cleared the tower'), flight control was handed over from the Kennedy Space Center to the Johnson Space Center in Houston (as it was in the days of Apollo). Shortly afterward, the shuttle started the 'roll-and-pitch programme' so that the orbiter flew upside down and at an angle to the vertical. The roll gave the pilots a clear view of the horizon (otherwise obscured by the external tank), so if they needed to abort and glide back to Kennedy or another landing strip, they could orient themselves immediately. The pitch was intended to ensure that the force of gravity, acting downward on the shuttle as it climbed, pulled the flight path more and more nearly parallel to the Earth's surface.
- At about 50 seconds into the flight, the speed of the craft built up to 1,190 km per hour and the altitude reached 9.1 km. As the shuttle made the transition from moving slowly through denser air, to moving quickly through much thinner air, the onboard computers throttled back the main engines (the SRBs also reduced thrust due to the design of the internal propellant moulding) to ease the craft through this period of maximum dynamic stress. Then the engines were throttled back up again.
- Just after the 2-minute mark, the SRBs were jettisoned. They continued to climb to about 67 km altitude, then parachuted down to the sea about 230 km from the launch site.
- An abort return to Kennedy became impossible 4 minutes and 30 seconds into the flight due to the shuttle's speed, altitude and distance from the launch pad and other sites became potential targets.[6]
- Within 6 minutes of launch, the shuttle was travelling at 20,000 km per hour, completely upside down and parallel to the Earth's surface. However, that speed was not enough to maintain an orbit. The engines continued to burn as the shuttle started a slow 2-minute descent back to Earth. This manoeuvre enabled it to increase velocity, aided by the pull of gravity. During this phase, the main engines gradually throttled back. The external tank was almost empty, so if the engines continued at maximum thrust, the much-reduced mass of the shuttle would cause the acceleration to build beyond the design limit of the structure. Acceleration was kept to a maximum of 3g.
- When the main engines powered down, 8 minutes and 30 seconds after launch, the shuttle was in an elliptical orbit with a perigee of 65 km and an apogee of 296 km above the Earth's surface. Twenty seconds later, the explosive bolts holding the external tank in place fired, and it fell back to Earth burning up in the upper atmosphere.
- The orbit was still too low to be maintained. Friction with the upper atmosphere at perigee would have eventually caused the craft to re-enter. For this reason, as the orbiter reached apogee (about 50 minutes after the main engines cut out) the Orbital Manoeuvring System engines fired, producing a $\Delta V = 240$ km per hour, circularising the orbit at 296 km altitude.[7] It might seem easier to burn the main

engines longer and inject the orbiter directly into this orbit, but that would have required a larger external tank to carry the extra propellant, making it harder to accelerate the stack off the ground. The technique used was more efficient. Controlling where the external tank burnt up would also have been much harder from the higher final orbit.

8.7 CHALLENGER

For a successful technology, reality must take precedence over public relations, for nature cannot be fooled.

Richard Feynman
Appendix to Rogers Commission Report

The morning of January 28, 1986, was cold: NASA had never launched a space shuttle in such low temperatures before (2°C, a full 15°C lower than the previous coldest launch, Figure 8.13).

At 11:38 Eastern Standard Time, the shuttle lifted off on the tenth flight of the *Challenger* orbiter and the start STS-51-L. Seventy-three seconds later, an explosion ripped the external tank apart, threw the SRBs away to steer drunkenly across the sky and blasted the orbiter into pieces. Most people viewing the

Figure 8.13 The morning of the *Challenger* launch was unseasonably cold. Ice can be seen in the vicinity of the SRB which is visible behind the scaffold along with the orbiter's wing. (Image courtesy of NASA.)

scene assumed that the seven crew members were killed instantly in the huge explosion. Some saw the SRBs descending in the distance on their parachutes and thought that the crew had escaped.

In fact, analysis of the film shot of the launch clearly showed that the flight and mid decks of the orbiter had been thrown clear of the explosion and fell 14 km to the sea amid a rain of debris. Examination of the recovered wreckage showed that the crew compartments were virtually unharmed by the explosion. The impact with the water at 333 km per hour (an estimated deceleration of ~200g) created the damage seen. The exact time at which the crew died is not known.

Certainly, the impact with the water was not survivable, but there is debate as to how many, if any, crew members were conscious and active during the fall. Three Personal Egress Air Packs[8] found on the flight deck had been activated, presumably after the breakup. The air remaining in the packs suggested that they were in use for a period consistent with the 2 minute and 45 second fall of the crew decks. If the compartment had been breached, the crew would have been unconscious within a few seconds at that altitude (the personal packs supplied unpressurised air). However, if the pressure integrity had been maintained, or degraded slowly, the crew may have been active for longer. The lack of buckling to the mid deck floor suggested that rapid decompression did not occur. The investigators found that some of the controls had been changed from their standard configurations, which could not have happened during the impact.

US President Ronald Reagan appointed former Secretary of State William Rogers to chair a committee to investigate the accident. Among the members selected were Neil Armstrong and the famous theoretical physicist Richard Feynman. All shuttle flights were postponed until the investigation determined the cause of the tragedy. The next shuttle mission did not fly until September 29, 1988.

8.7.1 PHYSICAL CAUSES OF THE DISASTER

The key events in the launch sequence leading up to the disaster were as follows.

- Photographs taken by cameras arranged around the launch pad showed that 0.678 seconds after the shuttle lifted off, the first of nine puffs of grey smoke appeared near one of the field joints at the bottom of the right-hand SRB (Figure 8.14). The smoke emerged with a frequency of about 4 times per second until 2.5 s after launch. This roughly matched the frequency at which loads were placed on the SRB structure by the launch, and hence the way in which the joint flexed.

A field joint consisted of a *tang* (male part) at the bottom of the higher segment and a *clevis* (female part) at the top of the lower segment. The small remaining gap between the joints was sealed by a pair of rubber O rings ~6 mm thick running around the circumference of the booster. The physical properties of the rubber used to make the rings was vital, as they had to expand with the joints to maintain the seal under load.

Analysis by the engineers of the Morton Thiokol Company responsible for the joint seals showed that during the launch, the pressure of the exhaust gases caused the casing of the SRB to bulge outward between the joints. This effect did not occur in the joint area because the casing was naturally thicker there. However, the bulge opened a gap between the tang and clevis (known as 'joint rotation'). The engineers observed on previous flights that the O rings were occasionally blackened and burnt by gases leaking through this gap. On the morning STS-51-L launched, Morton Thiokol engineers recommended a postponement as they were worried that the intense cold would prevent the rubber of the O rings from expanding quickly enough to seal the gap when joint rotation occurred.

- About 37 s after launch, *Challenger* started to feel the effects of high-altitude winds pushing it off course. This continued until ~64 s into the flight. The onboard systems compensated for the winds and the SRBs altered thrust direction as instructed. Conditions were such that the system was more active than on any previous flight.
- As the shuttle accelerated to the speed at which the atmospheric drag on the ship would peak, the main engines and SRBs reduced thrust to smooth the transition into the cleaner airflow beyond this speed. After that, the engines were throttled up to 104% of nominal thrust. Computer enhancement of launch images later revealed a small flame 58.788 s into the flight. The flame was near the joint from which the smoke issued. Presumably, whatever re-sealed the joint earlier had now failed.

Figure 8.14 Evidence of an unsealed joint: smoke emitted from the right hand SRB. (Image courtesy of NASA.)

- The flame can be seen without image enhancement in the next frame. By 59.262 s, it was a continuous plume. Data relayed to the ground by onboard sensors at 60 s into the flight showed a difference in internal pressure between the right and left SRBs, indicated that a leak was taking place.
- As the plume grew, the airflow over the shuttle deflected it so that it played onto one of the lower struts between the SRB and the external tank and onto the external tank casing (Figure 8.15).
- At 64.660 s, the flame changed shape and colour, indicating that it had burnt a hole in the external tank and mixed with the leaking hydrogen. Telemetry also recorded a drop in the helium tank pressure, confirming a leak. Some 45 ms later, a bright reflection appeared on the black tiles on the underside of the orbiter.
- Some 72.20 s into the flight, the lower SRB support structure failed and the right-hand SRB rotated downward about its upper strut. The subsequent impact dented and burnt *Challenger*'s wing.
- At 73.124 s, a pattern of white vapour developed around the edge of the lower dome of the external tank. Shortly afterward, the entire aft dome fell away, releasing massive amounts of liquid hydrogen. This created a sudden forward thrust of about 12,000,000 N, forcing the remains of the hydrogen tank upward towards the LOX tank. At the same moment, the rotating SRB struck the intertank structure and the lower part of the LOX tank. Both failed at 73.137 s. Within a few milliseconds, the liquid oxygen and hydrogen leaking from the tanks started to burn with almost explosive speed.
- As *Challenger* reached Mach 1.92 at an altitude of 14 km, it was enveloped in a burning cloud of propellant and the orbiter broke apart under the stress. The hypergolic propellants used in the Reaction Control Systems burst free and the reddish-brown colours they produced on burning can be seen at the edges of the main cloud. The film also shows several large pieces of the orbiter falling from the fireball. A wing, the tail section with the engines still burning and the forward fuselage trailing a mass of pipes and wires pulled from the payload bay can all be identified on the film (Figure 8.16).

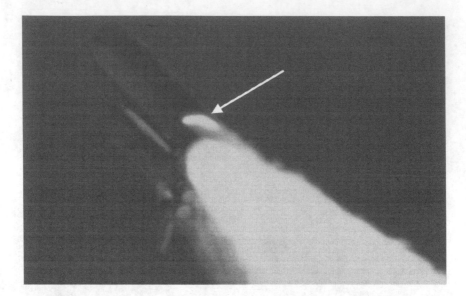

Figure 8.15 Exhaust from the failing SRB joint starts to play on the external tank. (Image courtesy of NASA with the author's addition.)

Figure 8.16 The final moments of the *Challenger*. The crew compartment can be seen among the falling debris. (Image courtesy of NASA with the author's addition of the ring.)

8.7.2 INVESTIGATION CONCLUSIONS

The investigators decided that the accident resulted from a failure of one of the field joints on the right-hand SRB. The problem of improper sealing under joint rotation was exaggerated that morning by the very cold conditions that stiffened the rubber in the O rings so they could not expand fast enough to maintain the seal.

In an appendix to the report, Feynman stressed the importance of listening to the engineers responsible for maintaining the shuttle, who urged caution in regard to several areas of operation. He was especially scathing regarding the process used to estimate the likelihood of component failure. The official figures cited the chance of a severe problem as 1 in 100,000 launches, meaning that a shuttle could be launched every day for 300 years without a major issue. More junior engineers estimated the odds as around one in a few hundred.

8.8 COLUMBIA

This is a bad day. I'm glad that I work and live in a country where we have – when we have a bad day, we go fix it.

Milt Heflin
Chief Flight Director, February 1, 2003

In essence, *Columbia* was lost on re-entry due to a breach in the Thermal Protection System (TPS) during launch. A piece of thermal insulation foam broke off the external tank some 82 s into the launch sequence and struck the left wing of the orbiter near the area where the leading edge curves away from the main body (Figure 8.17). The impact damaged the Reinforced Carbon–Carbon (RCC) panels on the lower leading edge of the wing.

In the intense heat of re-entry, the damage became a breach that allowed hot gas to penetrate the left wing and burn its way through the structure towards the landing gear. Mission control started to receive telemetry indicating an issue with the orbiter as various sensors in the wing failed. Meanwhile, the orbiter's autopilot was adjusting to changed aerodynamic conditions on its left side by firing the reaction jets.

Eventually, penetration of the gas into the wing severed all three redundant hydraulic lines and the orbiter lost the ability to control its aerodynamic surfaces and hence its flight path. Within a minute of this event, the crew was most likely killed (or at least rendered unconscious) by depressurisation of the crew compartment. The grim task of analysis after the accident revealed the effects of rapid depressurisation on the crew remains. The recovered suit helmets still had their visors in the open position. Closing and locking

Figure 8.17 *Columbia* on the launch pad prior to STS-107. Left: The vertical foam ramp covering the bipod support is shown inside the ring. The horizontal strut leading from the tank to the orbiter is also visible. Right: The ring indicates the region where the foam impacted the left wing. Most of the impact was on the underside of the leading edge. (Image courtesy of NASA; the two rings were added by the author.)

a visor in the event of depressurisation can be done rapidly. The fact that the visors were open indicates the crew must have been rendered incapable within a very few seconds.

The forward module broke free of the main orbiter body 31 s after loss of control. Four seconds later, the top and bottom parts detached from the pressure module, exposing it to the extreme conditions of re-entry. It is likely that the pressure breaches arose from the development of small holds in the mid deck floor area before the more catastrophic events that followed. At the time that the forward module separated, the rapidly decomposing craft was travelling at about 15 times the speed of sound at an altitude of 49 km. Only 35 *seconds* elapsed between the loss of aerodynamic control and the complete breakup of the crew module. It took about 35 *minutes* for the crew remains to reach the ground.

8.8.1 THE MISSION

STS-107 was designated a science flight in June 1997 and initially scheduled for launch in mid-2000. In fact, it was delayed 13 times over 2 years for a variety of reasons, but primarily because other missions had priority. For STS-107, *Columbia*'s payload bay was configured with:

- the *Spacehab* double science module that provided working space and scientific equipment for the astronauts (this was to be its first flight);
- an interconnecting tunnel linking the mid deck with *Spacehab*;
- a pallet of six experimental packages collectively known as *Freestar*[9];
- the *EDO*[10] pallet with extra liquid oxygen and liquid hydrogen to power the orbiter's fuel cells for the 16-day mission.

During their 16 days in orbit, the crew:

- took measurements of the ozone layer;
- studied atmospheric aerosols over the Mediterranean Sea and the Sahara Desert;
- performed a variety of studies related to combustion including the production of the weakest flame ever produced (50 times less powerful than a birthday cake candle);
- performed some experiments to determine the human body's response to microgravity: respiration during exercise, protein manufacture, bone and calcium production, renal stone formation and changes in saliva and urine due to virus infections;
- took the first images of upper-atmosphere 'sprites' and 'elves' – the strange phenomena associated with intense electromagnetic impulses generated by lightning in storms near the surface;
- captured extremely rare images of meteoroids entering the atmosphere;
- conducted research on plant growth in microgravity;
- grew samples of bone and prostate cancer cells for research purposes;
- hatched a fish in an aquatic environment and a moth from its cocoon in continuing research on the effects of microgravity on organisms.

8.8.2 LAUNCH

The launch countdown lasted 24 hour hours longer than normal. The extension was planned to allow more time for fuelling the EDO pallet and the final stowing of plants, insects and other items for the science experiments. Launch preparations proceeded normally, culminating in solid rocket booster ignition at 10:39 am on January 16, 2003. The SRBs shut down and separated from the external tank 2 minutes and 7 seconds later. They were recovered from their splashdown in the Atlantic and refurbished for a later mission. The main engines continued to burn until 8.5 minutes after launch. Shortly after the main engine cut out, the external tank separated.

At 11:20 am, the orbiter used its OMS engines in a 2-minute burn to trim the orbit: 282 km above the surface with a 39 degree inclination to the equator.

8.8.3 FOAM STRIKE

The launch initially appeared routine and uneventful. Subsequent analysis showed that a combination of wind shear acting on the stack and a lower than normal burn rate from the right SRB caused both SRBs and the main engines to gimbal through greater angles than normally experienced, although they remained well within the design parameters at all times. The gusting wind also triggered back-and-forth sloshing of the liquid fuel in its tank, which in turn set the stack into slow oscillation during the flight. Such oscillations were not untypical in shuttle launches, although the amplitude was somewhat larger than normal in the case of STS-107.

Standard practice dictated that video footage and still photographs were taken of every space shuttle during launch. The image analysis team's first look at the available data from STS-107 took place 2 hours after liftoff and initially concluded that nothing of note occurred. However, after reviewing the high-resolution images available the next day, the team identified an event 81.9 seconds into launch in which a large portion of foam detached from the left bipod region of the external tank and struck the orbiter's left wing. The images also showed a shower of post-impact fragments (Figure 8.18). As none of the fragments passed over the top of the wing, the analysts concluded that the impact occurred on the lower part of the leading edge.

The shuttle's speed was 2,500 km per hour at the moment the foam broke free from the bipod area. As the foam detached, it had the same upward velocity as the shuttle. However, as the craft was within the atmosphere, the foam was instantly subjected to large frictional forces and rapidly slowed[11] to ∼1,600 km per hour, allowing the orbiter to catch up with it. In essence, the orbiter collided with the foam at ∼900 km per hour.

8.8.4 DEBRIS LOSS

In the immediate aftermath of the *Columbia* disaster, the Air Force Space Command reviewed its 3,180 radar and optical observations of the orbiter during the mission. The mass of data included observations of a small object shed from the spacecraft between 10:30 and 11:00 am on flight day 2. The object drifted

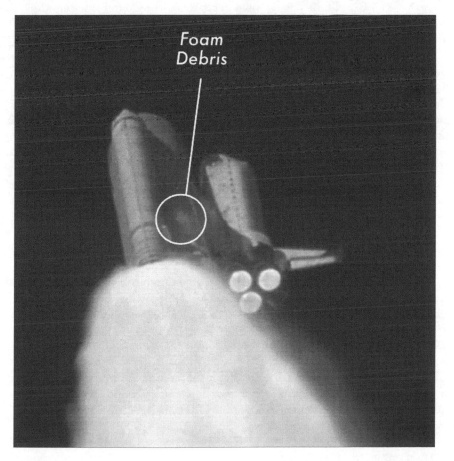

Figure 8.18 Foam debris scattering under the wing after a chunk of material was shed from the left bipod ramp. (Image courtesy of NASA's Columbia Accident Investigation Board.)

away at a speed between 1.1 and 5.5 km per hour and remained in orbit for ~60 hours before burning up in the atmosphere. Computer simulation of radar reflectivity and analysis of the object's orbital path confirmed that a piece of RCC panel was the best match for the characteristics of the tracked object.

The investigation board convened after the accident arranged for a series of tests, one of which involved firing a piece of foam at part of the leading-edge carbon protection removed from *Atlantis*. The panels had already flown 26 missions. When struck by foam travelling at 850 km per hour, the panel had a hole nearly 41×43 cm punched through it, leaving three large pieces inside the test wing. As the *Columbia* performed two manoeuvres shortly before the object was first tracked, it is possible that the yaw and attitude changes dislodged similar debris from where it settled inside the wing.

8.8.5 RE-ENTRY

At 8:10 am on February 1, 2003, the crew received a 'go' assessment for the de-orbit burn. This was carried out at 8:15 am with the orbiter flying upside-down and tail-first 282 km over the Indian Ocean. *Columbia* passed through the entry interface, the arbitrarily defined first contact with the Earth's atmosphere[12], at 8:44:09 am. Within 6 minutes, the temperature of the wings' leading edges would have reached 1,400°C.

A little over 4 minutes after entry interface, one of the sensors in the left wing recorded higher strains than previously encountered on *Columbia* re-entries. This information was logged in the onboard recorders but not transmitted to the ground and would not have appeared on the crew's display panel.

Columbia started a roll to the right to initiate a banked turn 323 seconds after entry interface. The manoeuvre was designed to manage the lift experienced by the craft and thus reduce the rate of descent to regulate heating. At this point, the ship would have been travelling at Mach 24.5. By 8:50:53 am, *Columbia*'s speed had reduced to Mach 24.1 at an altitude of 74 km and the craft entered 10 minutes of peak heating.

As the orbiter crossed over California, it appeared to people on the ground as a bright spot moving rapidly across the sky. As was normal for shuttle re-entries, many spectators gathered to watch and photograph the spacecraft's passage. The first sign that *Columbia* was shedding debris came at 8:53:46 when the orbiter suddenly brightened and a notable streak appeared in its luminescent trail. Spectators observed another four such events within the next 23 seconds, and eighteen more occurred over the next 4 minutes. The first indications of abnormality arrived at mission control at 8:54:24. One of the flight officers reported that four hydraulic sensors in the left wing displayed fault codes.

Columbia reversed its roll by turning right over Arizona at 8:56:30. It was travelling at Mach 20.9 and 67 km altitude. Within 2 minutes, as the craft crossed over New Mexico into Texas, a thermal protection tile broke away from the upper surface of the left wing. It was the most westerly piece of debris recovered.

The next fault indication came at 8:59:15, when pressure readings for both tyres in the left landing gear were lost. Shortly before this, the crew would have been alerted to the issue by an alarm on the flight deck. The commander and the pilot evidently called up fault pages for the messages displayed and attempted to diagnose the situation. Further indications came when the systems showed that the left landing gear was down but that the door over the gear well remained closed. The crew would have seen the messages as an instrumentation failure. The astronauts contacted mission control. The ground team's response was that they also saw the alarms but could not understand what the crew said.

At 8:59:32 am, mission control received a last verbal response from the shuttle: 'Roger …[word cut off]'. At the same time, the telemetry signal was lost. This was not unexpected as the timing coincided with a short-duration loss of signal as the orbiter switched antennas to relay data to the ETDRS[13] satellite and mission control.

Post-flight analysis and video imagery shot by onlookers along *Columbia*'s ground track established that a catastrophic event occurred at 9:00:18 am.

Mission control had no idea that it lost the orbiter and still expected to regain communications. By 9:02:00, they understood that the communications loss had continued far longer than expected, but still had no way to know what spectators across Texas saw: the orbiter was breaking up and debris was falling to the ground.

At 9:08:25, the team were trying a variety of measures to restore communications while attempting to establish the reason for the sensor loss that occurred before the break-off in telemetry. About 4 minutes later, *Columbia* should have banked to line up with Runway 33 at the Kennedy Space Center for its scheduled touchdown at 9:16 am. Ironically, around 9:12 one of the flight controllers received a call on his mobile phone telling him that live television coverage was revealing to America what the mission controllers still did not know: *Columbia* was lost during re-entry.

Shortly after that call was reported, the Flight Director issued the order to lock the doors to mission control. Within a few minutes of *Columbia*'s scheduled landing, NASA declared a 'shuttle contingency' and executed the plan put in place after the *Challenger* disaster. An investigation board was convened and a chairman appointed within an hour.

8.8.6 THE INVESTIGATION BOARD

In the immediate aftermath of the disaster, NASA's priority was the preservation of all evidence that could aid an investigation. While sealed in mission control, the flight team archived all their data and telemetry. At the Kennedy Space Center, various mission facilities and hardware, including Launch Complex 39-A, were put under guard or stored in secure warehouses. Similar steps were taken at other NASA and contractor facilities. It was clear from the start that the most crucial evidence was to be found in the rain of debris falling over Texas (Figure 8.19).

Cattle stampedes occurred in Eastern Nacogdoches County due to the sonic booms that were caused when pieces of the shuttle crashed through the atmosphere at 19 km per hour. A fisherman on Toledo Bend Reservoir watched a fragment hit the water and an unfortunate woman nearly lost control of her car when

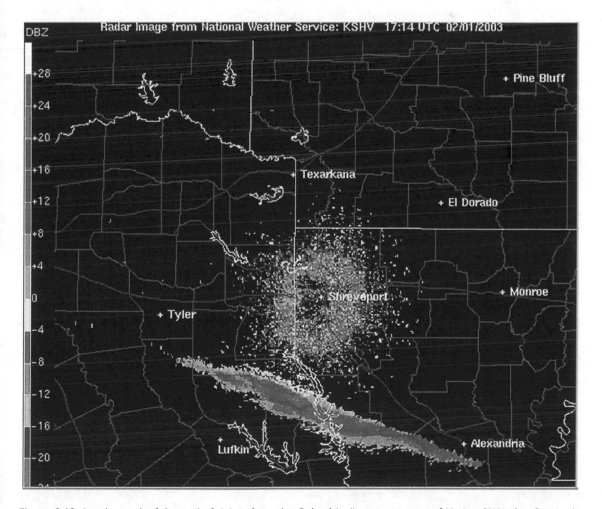

Figure 8.19 A radar track of the trail of debris from the *Columbia*. (Image courtesy of National Weather Service.)

part of *Columbia* hit her windscreen. Across Texas, 911 dispatchers were flooded with calls reporting sonic booms and burning wreckage.

What started as an emergency response soon grew into a massive undertaking with the need to decontaminate[14] and recover every fragment. By afternoon, reports arrived at a rate of 18 per minute. It was remarkable that little property damage occurred and no one was injured by tens of thousands of falling pieces.

In the end, more than 25,000 people participated in debris recovery and spent more than 1.5 million man-hours over an area larger than 9,300 km². Foot searches covered 2,800 km² and recovered more than 84,000 pieces, which accounted for 38% of the orbiter's unfuelled mass. A crucial data system recorder was recovered and provided temperature sensor readings from the left wing. Investigators were able to reconstruct the left wing's leading edge, enabling them to conclude that hot gases entered the wing in the region corresponding to the projected foam impact point determined by the photographic analysis of the launch.

8.8.7 FOAM LOSS DURING LAUNCH

While it proved impossible to establish exactly why the foam broke free of the external tank, the accident investigation board decided that several factors probably contributed:

1. At the time of the launch, foam loss had occurred in over 80% of the total 79 missions for which appropriate imagery of the external tank was available. Furthermore, in 7 of 72 missions for which imagery was available, the foam was shed from the bipod ramp.
2. Analysis of three un-flown external tanks revealed defects of various types (debris, voids and pockets) under the surfaces of the foam ramps. The ramps were hand-sprayed in thin layers to build volume and then carved to shape. Apparently, defects could build up along the lines between layers. This was caused by limitations of the process, given the properties of the foam, not from poor workmanship.
3. Nitrogen gas seeping from the intertank region could collect in voids in the foam and then liquefy. Drops of liquid that made contact with the very cold liquid hydrogen tank could freeze, only to 'flash evaporate' during launch, which might help to break away a piece of foam.
4. The design processes in the 1970s were limited by the existing technology and best engineering practices available. The optimal solution found for the location and structural design of the bipod support posed significant problems for foam application.

In essence, the investigation revealed that the design of the bipod support, a lack of understanding of the reason for foam shedding during launch, a possibly inadequate method for applying the foam and the extreme conditions experienced during launch all contributed to the *Columbia* disaster.

The engineers responsible for the space shuttle's initial design were very concerned about potential damage to the orbiter's fragile thermal protection system. The specific issue of foam loss during launch was cited along with concern about the possibility of other impacts. The *Flight and Ground System Specification Book 1: Requirements* states:

> **3.2.1.2.14 Debris Prevention**: The space shuttle system, including the ground systems, shall be designed to preclude the shedding of ice and/or other debris from the shuttle elements during prelaunch and flight operations that would jeopardize the flight crew, vehicle, mission success, or would adversely impact turnaround operations.
>
> **3.2.1.1.17 External Tank Debris Limits**: No debris shall emanate from the critical zone of the external tank on the launch pad or during ascent except for such material which may result from normal thermal protection system recession due to ascent heating.

Despite the clear design directive that the external tank *not* shed material, *Columbia* was showered with debris during its maiden launch, to the extent that more than 300 tiles had to be replaced after the landing. The engineers said that they would have expressed serious concerns about launching if they had known that the external tank would rain debris on the orbiter to that extent.

The space shuttle

The accident investigation board were very interested in how NASA's approach to foam loss drifted from extreme concern at the design and test flight stages to an 'in-family' issue[15] at the time of *Columbia*'s launch. Each successful landing despite damage due to foam loss and other impacts, increased NASA's confidence that foam loss was inevitable and not a flight safety issue. Such damage to the orbiter's thermal protection system was treated as a turn-around issue for the next flight, not a problem for a flight in progress. In blunt terms (and with the benefit of hindsight) nothing happened before the *Columbia* flight, so nothing was going to happen.

8.9 SPACE TRANSPORTATION SYSTEM (STS) EVALUATION

The goal we have set for ourselves is the reduction of the present costs of operating in space from the current figure of $1,000 a pound for a payload delivered in orbit by the Saturn V, down to a level of somewhere between $20 and $50 a pound. By so doing we can open up a whole new era of space exploration. Therefore, the challenge before this symposium and before all of us in the Air Force and NASA in the weeks and months ahead is to be sure that we can implement a system that is capable of doing just that.

George Mueller,[16] *opening remarks to the Space Shuttle Symposium, October 1969 – his target figure upper limit comes out, inflation-adjusted for 2011, to be* ~$680 *per kilogram*

...the Shuttle emerged from a series of political compromises that produced unreasonable expectations – even myths – about its performance ... in retrospect, the Shuttle's technically ambitious design resulted in an inherently vulnerable vehicle, the safe operation of which exceeded NASA's organisational capabilities as they existed at the time of the Columbia accident.

Columbia Accident Investigation Board final report, p21, August 2003

It is probably still too soon for the historical dust to have settled sufficiently for an objective and definitive evaluation of the space shuttle programme. However, several commentators and analysts have attempted to compare the shuttle's history with its original portrayal.

Looking back, it seems amazing that NASA originally proposed a turnaround time of 2 weeks between launches.[18] As the design progressed, it became clear that the shuttle being part of a more extensive Space Transportation System was unrealistic from a budgeting perspective and the focus shifted to just a reusable manned craft. Development was justified on economic grounds: if the one reusable vehicle could handle all government and private sector launches and service satellites in orbit, costs could be reduced dramatically. However, the turnaround capacity envisaged implied more flights than scientific and commercial needs seemed to warrant. Hence, NASA's argument hinged on persuading the Department of Defense to utilize the shuttle as the primary vehicle to deliver its payloads into orbit. The combined traffic from commercial, scientific and security payloads could have increased the number of flights per year up to the 50+ needed to achieve economic utility. Satisfying Department of Defense requirements, however, meant alterations to the design that added weight. The likelihood was that reconnaissance (spy) satellites would get larger and require bigger payload bays than those specified in the initial design studies. Also, the need to deliver payloads into a polar (rather than equatorial) orbit required more energy and so more powerful launch engines. The potential requirement for a landing at Vandenberg Air Force Base after a single polar orbit dictated a longer glide path and larger delta-shaped wings. It is probable that the adoption of solid rather than liquid propellant for the shuttle's reusable boosters was dictated on financial grounds.

A task for future historians is investigating how developing the Saturn V for payload delivery beyond Apollo may have impacted the shape of the space programme over the decades. Certainly, the shuttle absorbed a huge fraction of the NASA budget without ever reaching the flight turnaround times projected or reducing the costs of delivery to the anticipated levels.

NASA's cited cost for shuttle launches as of 2011 averaged $450 million per flight.[19] With a payload capacity of 27,500 kg to low Earth orbit (LEO),[20] this works out to $16,400 per kilogram. The Saturn V cost $185 million to launch[21] in 1971 (of which $110 million was vehicle cost). So, its 140,000 kg payload capacity to LEO represents $7,300 per kilogram when adjusted for inflation to 2011 figures. Clearly, such

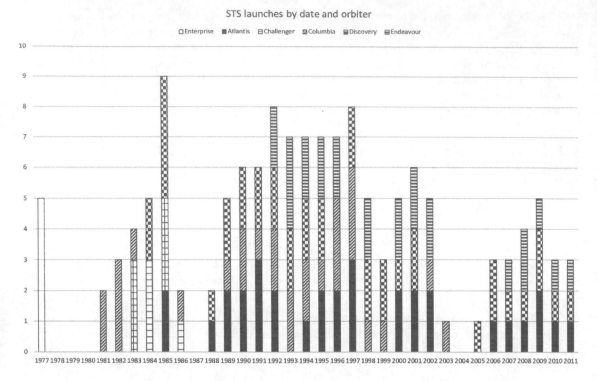

Figure 8.20 A chart of shuttle launches per year (1977 through 2011) by various orbiters. *Enterprise* was a test vehicle that conducted atmospheric flight and glide tests.

a naïve comparison does not tell the whole story but it certainly raises intriguing questions. We can see the outline of an argument to the effect that manned exploration of space has been restricted to LEO due to the costs of the shuttle and latterly the need to justify those costs as support for construction of the International Space Station (ISS). The impact of the shuttle budget on other developments also needs to be assessed. Surely no one in the 1970s would have expected the United States to have no manned launch capacity for 6 years (and counting) after the shuttle was retired.

Any retrospective view of the programme must include a consideration of the two heartbreaking disasters that occured, their human impact and the delays and cancellations that followed. Figure 8.20 shows the number of launches per year by all the orbiters from 1977 through 2011. The long periods when the orbiters were grounded after *Challenger* and *Columbia* were lost are evident.

Clearly, the exploration of space is a hazardous endeavour. Of 135 orbital launches, catastrophic shuttle failures occurred twice (1.5%). It is rare for failures of this scale to be attributed to a single engineering cause (the Apollo 13 incident is an example), but any objective critical analysis must pose questions about the complexity of the shuttle design, the level of compromise required and the human decision-making process at key times in the programme.

ENDNOTES

1. Various people suggested that rather than being allowed to burn up, the tanks could have been left in orbit and potentially marshalled together to construct an orbital hotel.
2. On the STS-71 mission to *Mir* in July 1995, seven people flew into orbit and eight returned – the largest number of people who ever flew in the shuttle.
3. These took place between December 1993 and May 2009.
4. There is a hierarchy of computer languages. Computer chips understand only very simple instructions and software written in their language is very complex and hard to follow. Computer languages that are of a higher level (such as BASIC, FORTRAN, C, VISUAL C++, etc.) are more easily understood by humans but have to be translated into machine language before the chips can run them.

5. Isn't HAL/S a curious choice of name for a language used to programme spacecraft computers?
6. NASA had visas on hand for immediate shipping to the appropriate country if the crew were forced to make an emergency landing.
7. These are the figures for a typical mission. Clearly, some have slightly different first orbits due to specific mission requirements.
8. Designed to provide the crew with oxygen should a mishap occur on the ground.
9. Freestar is the acronym for Fast Reaction Experiments Enabling Science Technology Applications and Research.
10. EDO is the acronym for Extended Duration Orbiter.
11. It took only 0.161 seconds for the foam to hit the wing.
12. An altitude of 400,000 ft.
13. East Tracking and Data Relay System. To that point, the orbiter communicated with the West Satellite.
14. The main concern was the poisonous nature of hypergolic propellants.
15. 'In-family' is an internal NASA term for a known flight issue that does not represent a threat to the shuttle and crew. The technical definition in the investigation board's report is 'A reportable problem that was previously experienced, analyzed, and understood. Out-of-limits performance or discrepancies that have been previously experienced may be considered as in-family when specifically approved by the Space Shuttle Program or design project.'
16. George E. Mueller (1918–2015) was a NASA associate administrator.
17. https://history.nasa.gov/SP-4221/ch6.htm
18. Production of external tanks by the contractors would have limited flights to about half that frequency.
19. https://www.nasa.gov/centers/kennedy/about/information/shuttle_faq.html
20. Low Earth orbit (LEO) is the altitude <2,000 km above the surface.
21. https://history.nasa.gov/SP-4221/ch6.htm

Intermission (5)

The International Space Station (ISS)

The idea of an orbiting construction in which human beings could live and work for extended periods has captured imaginations since the earliest thoughts about space travel. As a result, there is a long history of stations in Earth orbit that have met with varying degrees of success, culminating with the first permanently manned orbital outpost, the *International Space Station* (ISS) (Figure I5.1). Five space agencies are signed up to the ISS project:

1. NASA for United States;
2. Roscosmos for Russia;
3. JAXA for Japan;
4. CSA for Canada;
5. ESA for Europe (however, only Belgium, Denmark, France, Germany, Italy, Netherlands, Norway, Spain, Sweden and Switzerland participate in manned ventures).

The station has been continuously occupied since November 2000 and in that time 230 people have visited. It follows a near circular orbit some 410 km above the Earth's surface about every 93 minutes. The first component was launched on November 20, 1998 and the first residential mission arrived in November 2000. Assembly in-orbit has been from modules that have been robotically launched or carried into space by the space shuttle. Some 212 EVAs have gone into assembly and maintenance of the station—a total of 1,327 hours in space. Assembly continues and some components are scheduled for installation right up to 2022. The station is expected to be in service until 2030.

The current operational specifications call for a six-person crew and the station can accommodate six docked spacecraft at any one time. There are two sections to the ISS, the Russian Orbital Segment (ROS) and the United States Orbital Segment (USOS) which includes American, European and Japanese modules. In total, the pressurised internal volume of the ISS is comparable to that of a Boeing 747 jet.

The larger ISS modules are as follows (Figure I5.2).

- *Zarya*: The first module launched into space; this Russian component is also known as the Functional Cargo Block (FGB). It provided electrical power, storage, propulsion and guidance during the initial phases of assembly and now serves as a storage area.
- *Unity* (Node 1): The first American module was carried into orbit aboard the space shuttle in 1998. It is one of three nodes or connecting modules. Resources such as fluids, life support equipment, data

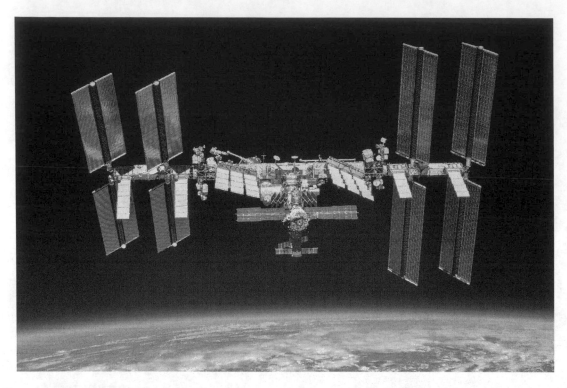

Figure I5.1 The International Space Station as photographed from a departing Soyuz in October 2018. (Image courtesy of NASA and the crew of Expedition 56.)

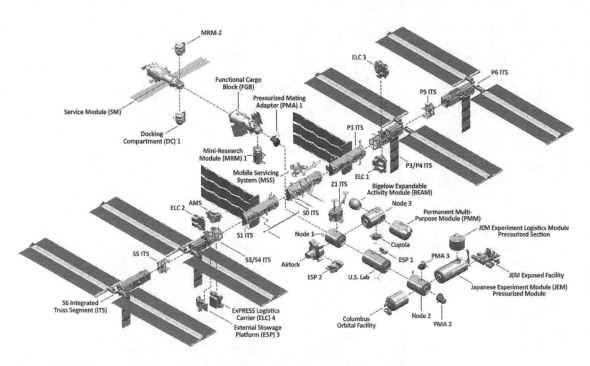

Figure I5.2 Elements of the International Space Station. (Image courtesy of NASA.)

systems and electrical power are routed through this node to supply other working and living sections of the station.

- *Zvezda* (DOS-8 or Service Module): Adding this Russian module made the station permanently habitable for the first time in the assembly sequence. It provides life support for up to six members and living quarters for two. As it is placed at the 'rear' of the station, judged from the direction of its orbit, this module's engines can be used to adjust the orbit.
- *Destiny*: The chief research module for the United States also serves as the primary mounting point for the truss structure that runs laterally across the station.
- *Harmony* (Node 2): This component acts as a utility hub for the USOS section of the station. It is connected to the *Columbus* and *Kibō* modules. Other ports on Node 2 are available as docking stations for visiting spacecraft. The USOS crew quarters are in this module including soundproof booths where crew members can listen to music, use laptops, and store personal effects.
- *Tranquillity* (Node 3): This node houses a water recycling system. Four of the node's six available ports are linked to other modules.
- *Columbus*: The primary European research module is equipped with a laboratory, and facilities for research in biology and fluid physics. Several experiments attached to the exterior deal with space science, Earth observation and technology.
- *Kibō* (Hope): This largest module is used for space medicine research, Earth observations, materials science, biotechnology and communications.
- *Cupola*: Attached to Node 3, this viewing area has seven windows including an 80 cm diameter round port – the largest window in space. The module contains workstations for operating the robotic arm attached to the station.

Among the regions of the station that do not require pressurisation, the largest and most distinctive is the Integrated Truss Structure (ITS) that runs transversely across the station and is 108 m long. Mounted on the truss are solar arrays in four wing-pairs that power the USOS. Each of these arrays is about 375 m^2. Rechargeable nickel–hydrogen batteries provide power for the 35 minutes of each orbit when the station passes into Earth's shadow. The Russian section derives its power from solar panels on *Zarya* and *Zvezda*. The other large panel-like structures running from the main truss are thermal radiators.

From the Moon to Mars

The NASA Administrator shall, 'Lead an innovative and sustainable program of exploration with commercial and international partners to enable human expansion across the solar system and to bring back to Earth new knowledge and opportunities. Beginning with missions beyond low-Earth orbit, the United States will lead the return of humans to the Moon for long-term exploration and utilization, followed by human missions to Mars and other destinations.

US President Donald J Trump
Space Policy Directive 1
December 11, 2017

9.1 FIFTY YEARS OF NASA (1969–2019)

On May 13, 2019, President Trump announced an amendment to his budget request that would add $1.6 billion to the $19.9 billion NASA budget already submitted for that fiscal year (February 2019). The proposal diverted money from other sources within the budget to support NASA's return to the Moon by 2024. The initial submission provided no funding for the International Space Station (ISS) after 2025, possibly seeking to extend funding from private means. It also directed NASA to expand commercial opportunities for transportation of crews and cargoes to low Earth orbit (LEO). In essence, the proposal would take LEO operations out of NASA's hands and focus the agency on space activities beyond LEO.

In response, NASA announced that *Artemis* (the mythological twin sister of Apollo and Goddess of the Moon) was to be the designation for the return-to-Moon project and reaffirmed its commitment to the ambitious timescale. However, the amount of money granted congressional approval remains to be seen.

President Trump's directive marks the latest chapter in the ongoing interactions between presidential administration, funding, congressional actions and NASA's plans and projects for space exploration. Since the end of Apollo, there have been several twists and turns, so we should assess the current plans in the light of past cancellations, technological developments and the focus within the agency.

9.1.1 THE POST-APOLLO YEARS

In the immediate post-Apollo years, there was little momentum carried forward from lunar exploration. President Nixon approved the development of the space shuttle in 1972, but NASA's budget fell steeply as a proportion of the total federal budget over the next 10 years. Despite this, NASA had ambitious aims: the establishment of a lunar base in the 1970s and a mission to Mars in 1981. Nixon did not sanction either plan but a 5-year cooperative arrangement with the USSR, culminating with the Apollo-Soyuz mission in 1975, was set in motion. The *Skylab* Space Station (developed from a Saturn V third stage) was established in 1973 and supported three crewed missions to 1974 (Figure 9.1).

After the Nixon presidency ended in disgrace, Gerald Ford (1974–1977) and then Jimmy Carter (1977–1981) followed without adding much impetus to space matters, at least as far as manned exploration was concerned. On the other hand, NASA enjoyed success with robotic missions.

Mariner 9 became the first probe to orbit another planet in 1971 and during its mission vastly increased our knowledge of the Martian surface. *Pioneer 10* was launched to Jupiter in 1972 and flew through the Jovian system in 1973. *Pioneer 11* followed in 1973, passing Jupiter in 1974 and Saturn in 1979. The two *Viking* probes were sent to Mars in 1975. Their landers touched down on the surface in 1976 and transmitted spectacular images back to Earth (Figure 9.2).

Voyager 1 set off for to Jupiter and Saturn in 1977. *Voyager 2* was a few weeks before *Voyager 1* but took longer to arrive at Jupiter and then moved on to Saturn, Uranus and finally Neptune. Amazingly, both probes

Figure 9.1 The *Skylab* Space Station as photographed by its last crew. Due to damage sustained during its launch, one of the solar panels was lost and the other jammed. The thermal insulation was also damaged. The first crew performed sufficient repairs to make the station operable. (Image courtesy of NASA and Skylab 4 crew.)

Figure 9.2 A Martian surface image sent back from the *Viking* I lander. Trenches dug into the regolith by the sampling arm can be seen. (Image courtesy of NASA and Roel van der Hoorn.)

remain in contact with ground control, more than 41 years after launch and being in interstellar space. Spectacular and scientifically important as all these missions were, their approval and design dated back to the 1960s, and they were scaled-back versions of more ambitious plans constrained by budgeting cuts.

In 1971, the Russians lost the crew of *Soyuz 11* when their capsule depressurised during module separation prior to landing. During the mission, the cosmonauts had entered the first space station *Salyut 1*. After that, the Russians launched and successfully occupied five more stations concluding with *Salyut 7* in 1982 (de-orbited in 1991). A mixture of civilian and military crews amassed a total of 1,696 days in orbit. Following the Apollo-Soyuz linkup (1975), NASA did not have another manned mission until 1981.

9.1.2 SPACE SHUTTLES AND SEI

The next president to drive the US space program forward was Ronald Reagan, who held office between 1981 and 1989. The first flight of the space shuttle (1981) took place early in his first term. He announced a range of space initiatives, including the construction of the space station (named *Freedom* at the time) and the Strategic Defense Initiative (nicknamed Star Wars) in 1982. Despite seeking to increase the NASA budget, many of his initiatives never became realities. NASA's reputation took a severe dent after the *Challenger* accident in 1986 and the publication of the Rogers Commission Report with its critique of NASA's internal culture.

In 1981, the USSR launched the first component of the *Mir* Space Station, which was completed in 1996 and remained operational until 2001. European astronauts started to visit Mir in 1988 a practice that continued until 1999.

On July 20, 1989, President George Bush, in an echo of President Kennedy's speech, commited America to landing a man on Mars. Twenty years to the day after the Apollo 11 landing and flanked by its three astronauts, President Bush launched a 10-year programme known as the *Space Exploration Initiative* (SEI), that would build up to a Mars mission via a space station and a return to the Moon. In response to the announcement, Richard Truly, NASA's administrator, commissioned a study (which came to be known as the '90-Day Study') of options available for achieving the SEI goals. The results were published in November 1989 and outlined plans for a modified space station (the ISS) and a lunar base (to be established in the early 2000s), both of which were specified as requirements to support an eventual mission to Mars. The payloads envisaged for these projects necessitated the development of a heavier-lift booster capacity.

The Mars mission plan proposed using nuclear thermal engines for the transfer and aerobraking into orbit at Mars. The journey time was 200 days with a surface duration of 600 days. The payload for LEO departure to Mars was estimated at 1,300 tonnes, with an additional 840 tonnes of propellant.

In hindsight, the report seems like a shopping list, seeking funding for a variety of internally supported 'pet' projects, rather than a clearly focussed plan for Mars exploration. The total estimated cost came to $550 billion (over 30 to 34 years) or $1.1 trillion adjusted to 2019. NASA's annual budget when the report was released was $11 billion: the SEI would have required twice that amount maintained over a span of about 30 years.

The *National Space Council* (NSC), created in 1989 and chaired by Vice President Dan Quayle, convened a special panel to review the 90-Day Study and consider alternative plans.

Many members of the panel criticised NASA for its unrealistic approach in the study, especially in the light of other, rather cheaper, alternatives that came to light during the review. The panel suggested that the *National Academy of Sciences* study the matter further. In May 1990, at the urging of the NSC, NASA created *Project Outreach* to review suggestions submitted from outside the agency. *Outreach* was chaired by Tom Stafford, the former Apollo astronaut, and their report was published in the summer of 1991.

In the meantime, NASA experienced more public relations setbacks. The main mirror of the *Hubble Space Telescope*, deployed in April 1990 from *Discovery*, was ground to the wrong shape. As a result, it produced images that were barely viewable rather than groundbreaking. *Discovery*'s launch had been the third of 1990, and NASA planned to make that year the most productive for shuttle missions since 1985. However, the next launch did not take place until October 1990 due to persistent and dangerous hydrogen leaks affecting the external tanks of more than one orbiter. The American press seized the opportunity to ask how NASA could conceivably fly to Mars if it could not produce a correctly shaped mirror or get the space shuttle off the ground.

Figure 9.3 The *Sojourner* rover from the *Pathfinder* mission analysing a rock the ground team named *Yogi*. (Image courtesy of NASA.)

In the light of these embarrassments, Vice President Quayle, under the urging of council members, called for an independent review of the space programme, which became the *Augustine Committee*. Its December 1990 report concluded that conditions within the agency were not as bad as they appeared to the public. However, the committee urged NASA to re-focus on science, curtail its attempts to formulate 30+ year plans and concentrate on incremental projects that could reasonably be expected to receive funding at each step. Rather noticeably, the committee did not come out strongly in favour of SEI.

Some years later, Vice President Quayle's memoirs revealed that the committee originally placed human exploration last on its list of suggested NASA missions. Under objections from Quayle and others, the listing was removed.

The following summer, Stafford's *Project Outreach* published its findings indicating that although the 90-Day Study had been unrealistic, there were no cheap ways to get to Mars and no radical technologies on the horizon that could significantly reduce costs. In essence, SEI was dead.

In 1992, Presidents Bush and Yeltsin announced the Agreement between the United States of America and the Russian Federation Concerning Cooperation in the Exploration and Use of Outer Space for Peaceful Purposes. The agreement triggered the Shuttle-Mir programme and US astronauts started having tours of duty on Mir. Mike Foale served on Mir as part of this initiative, but under less than ideal circumstances. During his tenure, one of the station's modules was struck by a resupply vessel, depressurising the module and sending the station into a spin. The crew eventually corrected the situation and the wounded station continued in service.

9.1.3 FASTER-BETTER-CHEAPER

In spring 1992, NASA Administrator Truly was replaced with Dan Goldin, who advocated a 'faster, better, cheaper' approach that led to a substantial reform of NASA's robotic spaceflight effort. In this context, manned exploration beyond LEO was unaffordable. Golding worked on gaining a commitment from the newly in office President Clinton to secure the development of the space station and to bring Russia into that programme. In September 1993, Vice President Gore and Prime Minister Chernomyrdin announced plans for what became the International Space Station (ISS).

The National Space Council was disbanded in 1993 and absorbed into the new National Science and Technology Council formed under an executive order.

NASA was not yet finished with public embarrassments, however. In the fall of 1993, NASA lost contact with its $813 million *Mars Observer* probe 3 days before it was due to enter Martian orbit. The mission was intended to study the planet's surface, atmosphere, climate and magnetic field.

A subsequent review in January 1994 concluded that the spacecraft had probably been disabled due to a leak that allowed the hypergolic propellant to mix prematurely outside the combustion chamber. Goldin's 'faster, better, cheaper' approach mandated cost-capped and tightly focussed missions for the future. The goals were to reduce development times, reduce costs and increase scientific gains by flying more missions in less time. Although not widely known, one aspect of Goldin's approach was to reduce the earlier 'all-eggs-in-one-basket' missions such as *Mars Observer* and demonstrate that it was okay to fail (with cheaper missions that pushed technology).

The *Mars Pathfinder* was a classic example of 'faster, better, cheaper'. After launch on December 4, 1996, it landed on Mars in July 1997. This highly successful mission featured the first Mars rover, which operated on the surface for 85 days, produced 550 images and analysed chemical compositions at 16 locations near the lander (Figure 9.3).

NASA operations during the Clinton administration revolved around shuttle missions, ISS construction and robotic exploration of the solar system. The ISS accepted its first long-term residents in November 2000 and has been manned continually ever since. The last pressurised module was added in 2011.

9.1.4 THE VISION

In the early days of George W Bush's time in office (2001–2009), NASA was again rocked by disaster when *Columbia* disintegrated on re-entry. The accident review board released its conclusions in August 2003, and in January 2004, the president announced the *Vision for Space Exploration*, soon followed by formation of the *President's Commission on Implementation of United States Space Exploration Policy*. The NASA *Exploration Systems Architecture Study* in mid 2005 developed the technical plans needed to implement the aims of the Vision. The goal was to develop a sustained programme of robotic and human exploration that was affordable and would recapture public enthusiasm for space matters. The scope of the study included planning beyond the completion of the ISS and the retirement of all space shuttles in 2010, and a return to the Moon by 2020, in preparation for missions to Mars and potentially beyond.

As a result, NASA embarked on the *Constellation* programme which included the *Orion* crew module, a lunar lander called *Altair* and the *Ares I* and *Ares V* boosters. A primary goal for Constellation was to develop manned systems and boosters to replace the space shuttle. The *Orion Crew Exploration Vehicle* was intended for use to LEO. Lockheed-Martin was named the prime contractor in August 2006 and a secondary contract, to construct a heat shield, went to Boeing. The Block I Orion was slated for ISS missions with Block II and Block III variants in mind for deep space flights.

The Ares I consisted of a single solid rocket booster (SRB) derived from the space shuttle design with a second stage fuelled by LOX and LH_2. The booster would have been able to carry Orion to LEO. Its development progressed up to a test launch in October 2009.

Ares V was more powerful, capable of delivering 188 tonnes to LEO (70 tonnes more than the Saturn V) and was not intended to carry crew. Its six engine first stage was supplemented by two SRBs that were 1.5 segments longer than those employed on the space shuttle. The second or *Earth Departure Stage* (EDS) was a development of the Saturn V third stage.

According to the mission plans for a lunar landing, Ares V would carry the EDS and Altair lander separately into orbit from Orion, which Ares I would lift into orbit. After Orion docked with the EDS/Altair stack, the EDS would inject the combination into a lunar trajectory, and then be jettisoned. On arrival on the Moon, Altair's descent engines would place the stack into an appropriate orbit for a polar landing. Polar targets were chosen as the most appropriate sites for sustainable Moon bases (see Section 1.4).

Unlike the Apollo missions, Orion would be capable of safely orbiting the Moon unmanned and monitored from Earth, allowing its crew of four to transfer to the lander and descend to the surface. The rest of the mission would proceed as did Apollo. The Altair ascent stage would re-dock with Orion, which would then return to Earth.

The peak of Constellation was to be a Mars mission in the 2030s. The outline plan[1] called for an Ares V launch carrying the Mars ascent vehicle (MAV) along with equipment for manufacturing propellant on the Martian surface (see Section 9.5.1.1). A second launch would carry the surface habitation module. Both vehicles would travel to Mars ahead of the crew (employing a nuclear thermal propulsion stage en route). The habitation module would go into Mars orbit and the MAV would land and convert atmospheric gases into propellant.

At the next Mars transfer window, the crew would journey onboard a nuclear–thermal interplanetary transfer vehicle, rendezvous with the habitat in Mars orbit and touch down near the fuelled MAV waiting for them on the surface.

As the Constellation programme continued, NASA initiated *Commercial Orbital Transportation Services* (COTS) in 2006, inviting companies to develop vehicles that could transport supplies and cargo to the ISS. In the words of the NASA administrator, Michael Griffin:[2]

> With the advent of the ISS, there will exist for the first time a strong, identifiable market for 'routine' transportation service to and from LEO, and that this will be only the first step in what will be a huge opportunity for truly commercial space enterprise. We believe that when we engage the engine of competition, these services will be provided in a more cost-effective fashion than when the government has to do it.

The fear was that NASA would not have sufficient funding to pursue the goals of the Vision without being able to deliver necessary supplies to the ISS in a more affordable manner.

After the shuttles retired, NASA initially relied on Russia and Europe for crew and material transport to the ISS. As a result of the COTS contracts, Space X and Orbital Sciences designed, built and flew spacecraft that delivered cargo by September 2013. Both companies were awarded contracts under the *Commercial Resupply Services* (CRS) umbrella in 2008 and various extensions (along with the addition of the Sierra Nevada Corporation in 2016) will take CRS through 2024. The COTS and CRS initiatives are evident successes. The $800 million invested in COTS produced two independent launch systems and vehicles, which contrasts markedly with Constellation's progress over the same period.

A 2009 report by the Government Accountability Office[3] quotes a NASA estimate that $49 billion of the $97 billion Constellation budget to 2020 was spent on the development of Ares I and Orion by that date. Both programmes faced technical challenges and solutions were not immediately forthcoming. As a result, it was difficult to project future costs.

9.1.5 MOON TO MARS

President Obama took office in January 2009. His federal budget request noted that Constellation[4] was 'over budget, behind schedule and lacking innovation'. The *Review of United States Human Space Flight Plans Committee* set up that year had a mandate to review NASA plans for human exploration after retirement of the space shuttle. In April 2010, Obama cancelled the Constellation programme along with plans to return to the Moon. Axing the lunar return phase was especially striking as NASA considered it an essential precursor to the asteroid and Mars explorations in planning.

The administration promised $3 billion as additional funding to develop a new booster programme with heavy lift in mind. The president called for the rocket to be ready for construction by 2015 and for manned missions to orbit Mars by the mid 2030s. NASA was directed to work with private companies to provide crew transports to the ISS. This was clearly seen as a way to drive costs down and push innovation after the success of COTS.

Despite the additional funding, the heavy lift vehicle was not ready for 2015 and Orion continued to face challenges. The new booster programme merged Ares I and Ares V development into the *Space Launch System* (SLS). The design passed a preliminary review in 2013 and the Block I version entered full development in 2014 with a projected launch in 2018. However, NASA announced in April 2017 that the launch would slip to 2019. Seven months later, a further delay pushed the launch back to 2020. Swelling development costs that outstripped NASA's available funding seem to be at the root of the problem.

On December 11, 2017, President Trump signed the Space Policy Directive-1, tasking NASA with landing astronauts on the Moon by 2024. In an earlier directive, the president re-established the National Space Council under the chair of Vice President Pence. Directive-1 calls for NASA to work with commercial partners to achieve a return to the Moon. Subsequently, in February 2019, NASA sought interest in developing three elements for Moon landings:[5]

- a reusable transfer element to carry crew and a lander from the *Gateway* (see below) to low lunar orbit;
- a reusable ascent stage capable of carrying crew;
- a descent stage.

Initially, NASA expects that only the transfer vehicle and ascent stage will be reusable. Developing the technology to extract propellant from lunar regolith and lunar water ice could allow the descent stage to refuel on the Moon and become reusable as well. Orion would deliver a crew to the proposed *Lunar Orbital Platform – Gateway* (Section 9.4) where they would board the transfer vehicle/lander combination for the flight to the Moon.

In November 2018, NASA announced the availability of *Commercial Lunar Payload Services* (CLPS) contracts in an effort to extend the philosophy behind COTS by using commercial systems to deliver science and technology payloads to the Moon in support of surface exploration. NASA has awarded contracts and selected 12 payloads for test delivery. A parallel call for out-of-agency payloads to the Moon for CLPS delivery closed in February 2019 with announcements to be made later in the year.

NASA continues to declare that exploitation of the Moon is an essential precursor to Mars exploration:[6]

> The Moon provides an opportunity to test new tools, instruments and equipment that could be used on Mars, including human habitats, life support systems, and technologies and practices that could help us build self-sustaining outposts away from Earth. Living on the Gateway for months at a time will also allow researchers to understand how the human body responds in a true deep space environment before committing to the years-long journey to Mars.

NASA clearly must accelerate plans and development to meet the presidential deadline. The watchword at the moment appears to be 'Moon to Mars', a motto that is displayed proudly on the agency's website.

9.2 ORION

For a return to manned launch capability, NASA is working with the European Space Agency (ESA) and Lockheed Martin to develop *Orion*, a successor to the Apollo CSM. Orion is intended to support missions as diverse as ISS crew rotation, lunar exploration and deep space missions to asteroids and eventually to Mars.

The roots of this design reach back to the cancelled Constellation programme which envisaged a somewhat leaner craft in three different versions to support diverse requirements. After Constellation was cancelled during the Obama presidency, Orion was redesigned as a multiple purpose vehicle. A test capsule was launched in 2014 using a Delta IV Heavy rocket[7] on an unmanned test flight that lasted 4 hours and 24 minutes, executed two orbits and splashed down successfully on target.

The next proposed test is to send another unmanned Orion around the Moon. The launch of this Exploratory Mission-1 (EM-1) is projected for 2020 using an existing launch vehicle.[8] The new SLS would then start delivering components of a lunar orbital station in 2021, leading the way for a manned lunar orbital mission in 2022.

9.2.1 THE SERVICE MODULE

In a significant step forward, the service module will use four independently gimbaled solar cells to provide electricity, rather than the fuel cells that Apollo employed (Figure 9.6). This saves a significant amount of size and weight, as not only are the fuel cells themselves not required, there is no need for large oxygen and hydrogen tanks to store the associated reactants.

Orion will employ a 120 V DC system with six batteries for power storage, whereas Apollo used three 28 V DC batteries. Orion's service module, like Apollo's, also contains thermal radiators and will be equipped with the normal thrusters for manoeuvring and docking purposes along with a 33,400 N main engine. For the first test flight and a couple more planned flights, this will be an Orbital Manoeuvre System engine left from the shuttle programme. Other designs for the permanent job are under consideration. The service module's dry mass of 6,185 kg compares with the 5,216 mass of Apollo.

9.2.2 THE CREW MODULE

Orion's crew module bears a striking similarity to the Apollo command module. It has the same blunt angular cone shape sloped at about the same angle as Apollo's. However, it is larger, ~3.4 tonnes less massive and built of the same aluminium—lithium alloy used for the space shuttle's external tank.

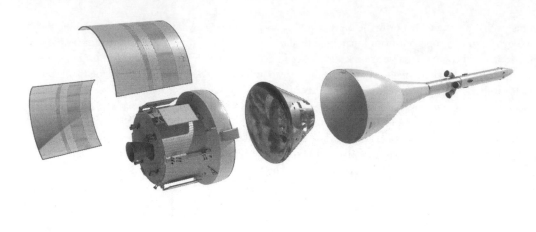

Figure 9.4 A rendering of the Orion crew and service modules, launch escape system and fairings that protect the delicate service module solar cells which are folded up next to the craft during launch. (Image courtesy of NASA.)

Figure 9.5 The European Space Agency–developed Orion service module during structural tests, November 2015. (Image courtesy of NASA.)

In another carry-forward, the upper external surfaces will be covered with Nomex thermal protection blankets. Orion is designed to be refurbished and reused but uses an ablative heat shield of the same honeycomb and resin construction as the Apollo command module's. However, as Orion is a significantly bigger capsule (3.3 m tall by 5 m across the base giving ~50% more internal volume than Apollo) its heat shield is the largest of its type ever constructed.

Figure 9.6 An artist's rendering of the Orion crew and service modules with the service module's solar array deployed. (Image courtesy of NASA.)

Orion has a nominal crew complement of four but in some situations (e.g. an ISS recovery mission) can be configured to carry six. The environment systems will provide a nitrogen–oxygen atmospheric mix at sea level or ~55% sea level pressure.

In another significant advance from Apollo, Orion will be controlled via a 'glass cockpit', derived from the design used in the Boeing 787 Dreamliner. Instrumentation will use screens instead of mechanical dials. Similar upgrades were made on the shuttle fleet starting in 2000. The new control panel will have only 60 switches nested around three large displays[9] (Figure 9.7). Engineers have been working with astronauts for 10 years on the designs of the hardware and software in simulations using mock-ups of the Orion cabin.

Figure 9.7 A mock-up of the Orion crew module at Johnson Space Center. The 'glass cockpit' display is visible above the astronauts. Note the two-by-two crew arrangement and the distortion of the control panel caused by the camera lens. (Image courtesy of NASA and Robert Markowitz.)

The ship's main computer is derived from a Honeywell design originally built for the Boeing 787, albeit with upgraded hardware to survive to rigours of launch and other manoeuvres. The system employs IBM PowerPC 750FX single-core processors, first manufactured in 2002. Although considerably less powerful than modern designs (they would be comfortably outstripped by modern smartphones), these chips are very reliable, especially in high-radiation environments.

Factored into the design of Orion is modularity. Components can be upgraded as new technology becomes available. This gives Orion a projected 30-year useable life.

Some crews will take Orion beyond low Earth orbit, and so outside the protection from radiation afforded by Earth's magnetic field. The same was true for the Apollo astronauts on the way to the Moon, but the projected Orion mission durations are much longer. During periods of intense solar activity when there is a high concentrations of ionising radiation in the solar wind, Orion's crew will use the aft bay in the capsule as a temporary shelter. Food, water, seating for launch and re-entry and other items positioned around the edges of the bay will provide shielding while utilising mass needed on-mission for other purposes.

9.2.3 THE LAUNCH ESCAPE SYSTEM

As with Apollo, one of the most hazardous periods of a mission is the launch. To maximise crew safety, a number of procedures must be designed that allow them to escape from the spacecraft at any phase, from waiting on the launch pad to achieving orbit. A critical component is the Launch Abort System (LAS) which may be used up to 120 s into a projected launch before jettison of the SRBs at 91 km altitude.

The design of the Orion LAS is superficially similar to that used on Apollo. A lightweight fairing fits over the crew module to protect it during flight through the thicker atmosphere. On top of this is a tubular rocket assembly containing three elements:

- Abort motors that can activate in milliseconds and pull the crew module and fairing clear of the booster.
- Attitude control motors to steer the capsule and fairing combination under command from the crew module.
- Jettison motors to pull the LAS clear, allowing the crew module's parachutes to deploy for a safe landing. The hypergolic fuel on the crew module would be dumped (at a safe altitude) prior to splashdown.

Only 14 s are required between LAS activation and jettison.

9.3 THE SPACE LAUNCH SYSTEM

Arising out of the cancelled Ares I and Ares V designs, the SLS is intended to be the most powerful rocket ever launched. Over time, it will develop into three versions with increasing capabilities and crew/cargo certifications (Figure 9.8).

Core stage

All SLS variants will employ the same core stage, 8.4 m in diameter, 65 m high and derived from the design of the shuttle's external tank. The stage will operate for up to 500 s and reach an altitude of 162 km. It employs four RS-25 engines. Initially these will be the RS-25D variety left from the shuttle programme. Ultimately NASA will switch to a more economical non-reusable version.

Block I and Block Ib

The Block I and Ib SLS will employ two SRBs derived from those designed for the space shuttle. Booster length has increased to 54 m via an additional centre segment for a total of five segments. Other aspects of the design such as avionics and insulation are updated. The new motors will provide 25% more specific impulse than the older versions and are not intended for reuse. They will fire for 126 s and take the SLS to ~90 km in conjunction with the core stage engines.

The Block I SLS employs an Interim Cryogenic Propulsion Stage (ICPS, 8.8 m × 4 m) carrying a single RL10 Aerojet Rocketdyne LOX/LH$_2$ engine with a carbon—carbon nozzle to improve the specific impulse. NASA's plan was to use the Block I design to launch EM-1, the unmanned Orion CSM mission to the Moon (see Section 9.3.1), although NASA's intentions may change under the pressure of time.

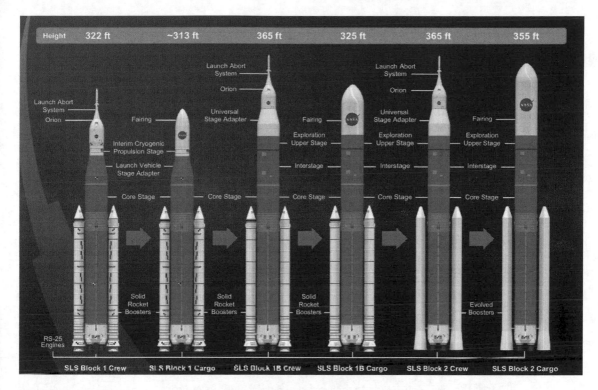

Figure 9.8 Projected versions of the space launch system. (Image courtesy of NASA.)

If the SLS is used, then the ICPS will inject the Orion onto its lunar trajectory. EM-1 will last 25.5 days including 6 days in lunar orbit. Thirteen CubeSats (miniature satellites) are to be deployed during the mission.

Block Ib replaces the interim stage with the Exploration Upper Stage (EUS). This 18 m long and 8.4 m diameter stage will employ four LOX/LH$_2$ engines delivering 440,000 N combined thrust. This stage replaces the earlier Earth Departure Stage (EDS) that would have been driven by three more powerful J-2X Aerojet Rocketdyne engines instead of the RL10 on the EUS. The change was made because the J-2X was not going to be ready in time. Further work on J-2X has been suspended until there is demand for transferring heavier payloads to Mars. Block Ib is currently intended for EM-2, the first crewed mission for Orion.

Block 2

Progress from Block I to Block 2 requires more advanced supplementary boosters to replace the SRBs. Several configurations are in competition:

- a LOX/RP-1 design using three engines per booster;
- an advanced solid-fuel booster using a composite casing rather than steel;
- a LOX/LH$_2$ design using two F-1B engines per booster.

The demand for Block 2 is not expected until the 2030s, so there is time for other technologies to enter into consideration.

9.3.1 CRITICISMS OF SLS AND CHANGING NASA OPINIONS

In October 2018, NASA's inspector general suggested that the development costs of SLS through 2021 would amount to about $9 billion.

A few months earlier, on February 6, Elon Musk's Space X launched *Falcon Heavy*, a reusable platform capable of delivering 64 tonnes to low Earth orbit at a claimed cost of $150 million per launch (the payload mass and cost are both for the rocket in its expendable configuration) or $2,351 per kg. The mission was an outstanding success and sent Musk's Tesla Roadster towards Mars, although it did not quite hit the

Figure 9.9 The synchronised landing of the two side boosters from Falcon Heavy's maiden flight. (Image courtesy of Space Exploration Technologies Corp.)

intended orbit. Both side boosters successfully landed back at Kennedy Space Center (Figure 9.9), but the attempted landing of the centre core failed when it did not re-light all its engines.

To date, there has been one further launch of Falcon Heavy, on April 11, 2019 delivered *Arabsat-6A* (a 6.5 tonne Lockheed Martin communications satellite owned by the Arab League). All cores of the rocket successfully landed.

Other commercial launches are planned through to 2022, apparently generating enough commercial income to cover the ~$500 million development costs. Space X's next phase of booster development (originally called the *Big Falcon Rocket* or BFR) will use a *Super Heavy* first stage and a second stage/spacecraft combination called *Starship*. This vehicle is projected to cost $5 billion to develop and will deliver 140 tonnes to LEO. The launch costs are not yet finalised.

It is hard not to contrast these launch and development costs, and rate of development, with the progress and budget of the SLS. As mentioned in Section 9.1.5, the SLS launch for EM-1 is delayed until 2020. During a Senate Commerce Committee meeting in March 2019, NASA Administrator Jim Bridenstine suggested[10] that SLS may slip even further behind. However, he remained committed to a 2020 date for EM-1 and indicated that the agency is looking at the commercial possibilities for the launch.

However, no commercial options exist to achieve this in a single launch. It would need two launches: one for the Orion command and service module and another for an upper stage. The module and upper stage would have to dock before proceeding to the Moon. At present, Orion has no docking capability, so something would have to be developed in time for the launches. Given the availability of launch vehicles such as Falcon Heavy or Delta-4, it is possible that planned launches might have to be bumped for EM-1.

There have been calls for the cancellation of SLS. Various bodies argue that the development programme is sapping funds from other NASA projects and will not reduce launch costs. Others debate the need for any heavy launch vehicle, since lower payload systems with capacities up to 60 tonnes exist and orbital assembly is certainly possible for lunar missions. Aside from the burgeoning development costs, it seems likely that the SLS will be expensive to launch, and so the budget may not exist to use it regularly.

The congressional mandates that forced NASA to use space shuttle components in the SLS design have also attracted criticism. An entire infrastructure of companies and jobs grew around the regular launches of the shuttle over its 30+ year history and Congress and the Administration did not want to see that disappear overnight. However, the result is that contracts have been awarded to existing suppliers without competition.

With SpaceX's stated aim of establishing a Martian base in the 2020s using Super Heavy and Starship the next decade or so will see an interesting comparison between the commercial exploitation of space, leading to commercial exploration, and the government-funded, politically supported approach that served NASA so well in the 1960s – if that same spirit can be restored and sustained against budget and political pressure.

9.4 LUNAR ORBITAL PLATFORM – GATEWAY

To support sustained lunar exploration and future missions to Mars, NASA intends to construct a space station in the vicinity of the Moon. This combination motel, fuel depot and logistics supply post would be a transfer point for crews to move from Orion capsules to lunar landers. The planned components of the Gateway (still in active development) are as follows (Figure 9.10):

- The PPE (Power and Propulsion Element) module will be the first to launch, probably via a commercial booster in 2022. This module will generate electricity for the station and contain its solar electric propulsion system. The 50 kW unit will handle station keeping and move the Gateway from one orbit to another, as required.
- ESPRIT (European System Providing Refuelling, Infrastructure and Telecommunications) will contain communications equipment, an airlock for science packages along with xenon and hydrazine storage capacity.
- Launched simultaneously with ESPRIT will be the US Utilisation Module, a smaller pressurised volume initially intended for food storage. The plan is that these two modules will be part of the second manned mission of the Orion capsule via the SLS projected for 2024 (EM-3). They would be linked to the PPE in a special orbit around the Moon (Section 9.4.1).
- The International Partner Habitat and the US Habitat are projected to launch in 2024 and 2025 respectively on the fourth and fifth Orion missions. They will provide total habitable volume of at least 125 m³ (ISS habitable volume stands at 332 m³).
- As of February 2019, Canada was the only partner to formally commit to the project and will supply the robotic arm. It is scheduled to be shipped to Gateway with the Gateway Logistics Module in 2024.
- The Gateway airlock module should be delivered in 2026 and will serve as a berth for docked spacecraft.

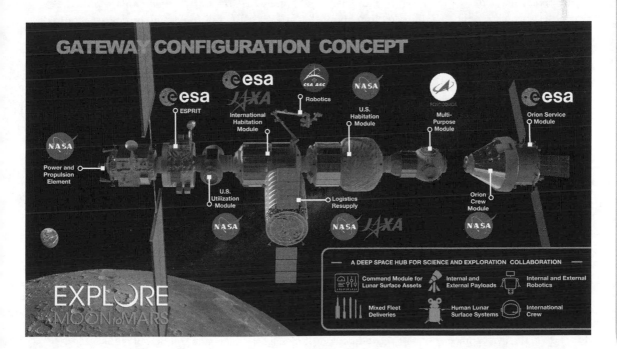

Figure 9.10 The proposed Gateway platform for lunar surface and deep space exploration. (Image courtesy of NASA.)

However, all of projected dates detailed above came from NASA documentation before Space Policy Directive 1 was issued. Some acceleration will be required in order to meet the presidential deadline.

9.4.1 HALO ORBITS

An orbit exists when the gravitational force acting on a body provides the necessary centrifugal force, given the radial size and speed of the orbit. When two massive bodies are involved, the options for a much smaller third body become somewhat more interesting.[11] There are five *Lagrange (L) points* in a two-body system in which an (in principle infinitesimal) third body can exist in a stable location relative to the others.

In Figure 9.11, Mass 1 is large compared to Mass 2 (like the Earth–Sun or the Earth–Moon combination), so that Mass 2 is essentially in orbit about Mass 1. Normally, an object at L_1 would orbit Mass 1 with a period rather shorter than that of Mass 2. However, that is to neglect the effect of Mass 2's gravitation on the situation. As the two large masses pull the smaller mass in opposite directions, its orbital period can be far shorter without it falling towards Mass 1. In fact, the whole point of L_1 is that the period becomes the same as that for Mass 2, so an object there is stationary relative to the larger masses.

Similar arguments apply to L_2 and L_3, except that the two large masses pull in the same direction at these points. The remaining L_4 and L_5 points are at the corners of equilateral triangles with side lengths equal to the Mass 1–Mass 2 orbital radius.[12] Unfortunately, $L_1 - L_3$ are unstable locations, in that a small disturbance moving an object off-station does not induce a restoring gravitational effect. The L_4 and L_5 points are stable, as a second-order effect will curve any trajectory taking an object off-station into a path around the point. Several planets have asteroidal debris (*Trojan satellites*) at the L_4 and L_5 points. Jupiter has more than a million such attendants.

The unstable nature of the Lagrange points has not prevented their exploitation for various purposes. *Halo orbits* circle the Lagrange points and can be maintained with minimal corrections. In 2018, China inserted a relay satellite named *Queqiao* into a vertical[13] halo orbit about the L_2 point in the Earth–Moon system. From this vantage point, the satellite kept the Earth and the far side of the Moon in continual sight, enabling the transmissions to and from the later (January 2019) *Chang'e 4* mission which soft-landed on the far side of the Moon.

The *Solar and Heliospheric Observatory* (SOHO, 1996) and *Advanced Composition Explorer* (ACE, 1997) solar observational satellites are in halo orbits about the Sun–Earth L_1 point. Although they are required to adjust their orbits from time to time, both have enough propellant left to operate into the 2020s. Also, NASA's plans for replacing the *Hubble Space Telescope* with the *James Webb Space Telescope* will place the new observatory in a halo orbit about the Sun–Earth L_2 point facing away from Earth, so that it does not have to contend with solar glare.

Most significantly, however, NASA's latest plan to locate the LOP-Gateway station is in a highly elliptical halo orbit about the Earth–Moon L_2 point. To a casual glance, it resembles a polar orbit around the Moon, with a closest approach at the lunar north pole of 1,500 km and an opposite point 70,000 km

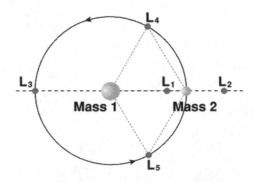

Figure 9.11 The five Lagrange points in a two-body gravitational system. (Image courtesy of NASA.)

from the surface. The elliptical halo is, in fact, an orbit about L_2 with a 6-day period. From this vantage point, the ΔV to a polar low lunar orbit is 730 m/s, a trajectory that would take half a day to execute. Like other halo orbits, this one is not completely stable, but keeping the station in orbit is projected to require a ΔV of only 10 m/s per year, so it is economical.

9.4.2 CRITICISM OF LUNAR GATEWAY

NASA justifies the establishment of the Gateway on the grounds that will provide experience and a technological test bed for eventual missions beyond the Moon. The agency hopes that long-duration occupation of the Gateway will support extended exploration of the Moon and determine the medical effects of working beyond LEO. According to NASA, sustained exploration of the Moon will present commercial opportunities and allow the establishment of a science outpost where planetary science, astrophysics, Earth observation, space biology and human health studies may be carried out.

Criticism of the plan has come from a variety of interested people and groups for diverging reasons. In November 2018, at a meeting of the National Space Council Users' Advisory Group, Buzz Aldrin voiced his opinion. He feels that a reusable lander in lunar orbit could achieve the same ends as the Gateway. At the same meeting, former NASA Administrator Mike Griffin (see Section 9.1.4) suggested that Gateway would become useful only after a lunar landing, not before. In his view, once the viability of using lunar materials to manufacture propellant was established, Gateway could serve as a refuelling depot serviced from the Moon, rather than from Earth. Robert Zubrin[14] (see Section 9.5.1) called Gateway 'NASA's worst plan yet' on the grounds that it is not needed and would not accomplish anything that could not be achieved (from a scientific and technological perspective) at the ISS. In his view, the most sensible place to build a Moonbase is on the Moon, where lunar material can provide resources and shielding from radiation never mind the scientific benefits of being on the surface.

These critiques, however, may miss a more fundamental point. The increased weight of Orion after its re-tasking as a multiple purpose capsule capable of supporting deep space missions means it cannot generate the ΔV required to move in and out of a suitable lunar orbit. In that light, the Gateway becomes an essential transfer point. Also, given NASA's current commitment to the SLS, assembling Gateway will justify that booster family. Gateway may also play a role in a future Mars mission.

9.5 ON TO MARS

NASA's plans for an eventual Mars mission leverage the Gateway and a propose a new vehicle, the *Deep Space Transport* (DST) or *Mars transit vehicle*. The DST will combine an Orion crew module and a habitation unit and utilise chemical and electrical propulsion. It is intended to support missions up to 1,000 days (2.8 years) and sustain a crew of four. The 8.4 m diameter habitation unit would be 11.7 m in length and constructed in the same manner as the SLS propellant tanks.

Although no direct request for funding appears in NASA budget submissions, the expectation is that the DST would launch to a rendezvous with Gateway in 2027 (EM-6). It would then undergo 100 to 300 days of crewed operation (EM-7) before departing on a one-year shakedown flight in lunar space around 2029 (EM-8). After that, the plan is for a manned orbital mission to Mars in 2033 (EM-11).[15] It is possible that the propulsion onboard the Gateway will be used to change its orbit to facilitate some of these missions, for example, by carrying Orion to the DST.

9.5.1 MARS DIRECT PROPOSAL

Dr Robert Zubrin is an American aerospace engineer and author. He is best known for his pioneering work on space exploration, especially the 1990 *Mars Direct* proposal popularised in the book *The Case for Mars*.[16]

Mars Direct came from a response to the SEI and 90-Day Study debacles. At the time, Zubrin worked at Martin Marietta (an aerospace company that merged with Lockheed in 1995 to form Lockheed Martin) along with David Baker. Zubrin saw the flaws in the SEI proposals for Mars missions. He felt that NASA's solution to deep space exploration, which involved throwing as much technology at the problem as possible, would escalate costs and be politically untenable.

The need to carry crew consumables (air, food, water) obviously constrains total mission time, unless the mass of the craft is increased, as a result of which more propellant is needed. For this reason, NASA's plans focussed on reducing journey time either via a Venus gravity boost (as discussed in Chapter 4), or a large ΔV trajectory, typically powered by a nuclear-thermal engine or chemical means requiring large amounts of propellant. The overall consumables budget can then be reduced if the first transit is coupled with a short surface stay.

To Zubrin and Baker, this design ignored the significant advantages of spending longer on the surface 'living off the land' and utilising the resources found on Mars. If you can find the means to support a crew for an extended surface stay, then a more conventional Hohmann-type transfer becomes possible by accepting the longer wait for a return window. Mars' atmosphere provides a degree of radiation shielding, which is hard to come by in space. Hence, day-for-day, the better scenario over a whole mission is to reduce the time in space compared to time on the surface. Additional shielding can be provided by 'digging in' (covering habitats with surface regolith) or even using underground tubes formed by lava flows as the basis of a habitat.[17]

In 1990 Zubrin and Baker worked out the substantial features of the Mars Direct plan. The breakthrough came out of a modification to the idea proposed by Jim French in an article for the *Journal of the British Interplanetary Society*. French suggested landing an automated propellant manufacturing plant on Mars before the crew arrived. The plant could then use local resources to produce stocks of propellant for the return journey. Since the manned mission would not need to bring propellant for the return on its outward bound trip, more supplies could be carried. The difficulty was landing the manned mission 'within a hose length' of the propellant plant. While Apollo 12 had proven the possibility of a precision landing, there is still a significant risk involved when you are relying on being able to refuel a lander on arrival.

In the Mars Direct version of the idea, the whole Earth return vehicle touches down (without crew) with an in-built plant to provide its own refueling. Having manufactured enough propellant for the return journey, the lander signals its readiness to Earth. The manned mission then departs, but in a much smaller ship, as the crew is not using it to return to Earth.

9.5.1.1 The Mars Direct profile

In detail, the Mars Direct plan works like this. The first launch sends an unmanned ship directly to Mars from Earth's surface (no assembly in orbit). This craft, with a mass of about 45 tonnes, follows a Hohmann-type transfer orbit and arrives on Mars by aerobraking to the surface. The craft carries supplies, equipment, a chemical plant and some liquid hydrogen.

Once on the surface it uses Mars' atmosphere to manufacture fuel for the return journey.

The atmosphere on Mars is a little less than 1% of the pressure at Earth's surface and composed almost entirely of carbon dioxide (CO_2). The lander reacts the liquid hydrogen with carbon dioxide from the atmosphere:

$$CO_2 + 4H_2 \rightarrow CH_4 + 2H_2O$$

producing methane (CH_4) and water (H_2O). The methane is stored as a liquid (perhaps by cooling via contact with the liquid hydrogen). The water is also stored and then decomposed electrically:

$$2H_2O \rightarrow 2H_2 + O_2$$

The resulting hydrogen is used to react with more carbon dioxide and the oxygen is liquefied and stored. Clearly, the oxygen is the oxidiser needed for the return journey and the methane is the fuel. Burning methane in oxygen at a mixture ratio of 2:1 produces a specific impulse of 340 sec. It would be better to use a ratio of 3.5:1 to generate a specific impulse ~380 sec, but that would require more oxygen than could be produced in the reactions detailed above. A number of possibilities, such as the direct decomposition of carbon dioxide, have the potential to produce the extra oxygen. Whatever process is chosen, more than enough oxygen can be generated to provide oxidiser for the fuel and also be stored along with water for the crew.

In 1993, Zubrin was part of a team that designed and built a working plant that could produce 400 kg of propellant for a robot probe designed to return samples of Martian soil to Earth. The plant had a mass of

20 kg, required 300 W of electrical power and functioned with 96% efficiency. There is little doubt that it could be scaled up for use in Mars Direct.

Having manufactured sufficient fuel and oxygen, the craft signals its readiness back to Earth. By using robots similar to the rovers that have wandered Mars for decades, the team on Earth explore the vicinity of the return vehicle, looking for a safe place for the crew to land. A radio beacon is then placed to guide astronauts to the selected site.

Once everything is ready for the next phase of the mission, a second return vehicle, identical to the one already on Mars, is dispatched on a 250-day transfer journey. A few weeks later, a third launch sends a manned version of the craft on a higher ΔV transfer orbit, so that the crew take about 180 days to reach Mars. This is more easily done in the lower mass craft that the crew will use. The ship's habitation module is a 5 m tall and 8 m diameter cylinder with two decks and enough supplies for an 800-day mission. The astronauts will know that a functioning and fuelled return vehicle is waiting for them on Mars. As a result, their craft needs only to be able to land on Mars. In this mission profile, there is no requirement for a 'mother ship', Apollo CSM style, in orbit with a crew member onboard, vulnerable to radiation, or sophisticated automated systems minding the house. The difficulties in landing a large vehicle on the surface that plagued the Direct Flight plan for the Moon will not apply. The crew will not return to Earth on the landing vehicle.

En route to Mars, the astronauts will disconnect their ship from the upper stage of the vehicle used for launch. Rather than ejecting the spent stage, they leave it attached to the ship by a long tether. The stage can then serve as a counterweight, allowing the combination to spin about its centre of gravity. This will generate some artificial gravity during the trip and reduce the health effects from long-term exposure to zero-g (see Section 4.7).

Upon arrival on Mars, the tethered upper stage is released and the lander aerobrakes to the surface within rovering distance of the return vehicle. Ample supplies on their craft and on the return vehicle will enable the crew to remain on the surface for 500 days until a low energy return window opens. They will have time to explore and perform scientific experiments. In the meantime, the second return vehicle will arrive and produce fuel. The next manned mission can start as soon as that vehicle is ready.

Mission design should always factor in a range of alternatives in the event a spacecraft develops a problem. The Apollo 13 mission turned out well because the crew survived in the lunar module, which was always considered a potential lifeboat. However, the Apollo crew's return to Earth covered far less distance than that facing a Mars expedition team. A Mars Direct mission would have several emergency options:

- When they are nearer to Mars than Earth, their best approach is to continue to Mars, where an inhabitable craft will await them, rather than trying to return to Earth.
- If they land further than a rover drive away from the return vehicle, they can guide the second return vehicle already en route to land nearby.

Mars Direct offers opportunities for sustained and systematic exploration of the surface by a sequence of such missions. It could all be done with a booster similar to the Saturn V.

9.5.2 NASA'S REACTION

Out of all the ideas that emerged during the SEI period, Mars Direct seems to have had the most lasting influence on NASA's plans and the thinking of mission enthusiasts around the world. In part, this is due to the inherent brilliance of the concept. However, as with the Lunar Orbit Rendezvous, having an enthusiastic and expert supporter with the energy and drive to push an idea is a huge help. LOR had the benefit of John Houbolt's expertise. It appears that Mars Direct benefited similarly from Robert Zubrin's efforts.

Zubrin and Baker presented their plans in a briefing to NASA engineers at the Marshall Space Flight Center in April 1990 and to the public in May at the National Space Society's conference. Over the summer, Zubrin made presentations at aerospace meetings and NASA conferences.

In 1992, NASA adopted a variant of Mars Direct in its baseline architecture for manned missions to Mars. A compromise proposal worked out by Zubrin and Dave Weaver (Johnson Space Center) known as Mars Semi-Direct is based on three launches per mission. One delivers an unmanned Mars ascent vehicle (MAV) to the Martian surface where it manufactures propellant. A second unmanned launch sends a fuelled Earth return vehicle (ERV) into orbit about Mars. The third launch transports the crew. They land near the MAV which has spent its time on Mars generating sufficient propellant for a return to orbit. At the end of the surface mission, the crew use the MAV to climb to orbit and dock with the ERV for the return to Earth, jettisoning the MAV in the process. The advantage of this profile over the full Mars Direct plan is the extra safety factor provided by the fuelled ERV in orbit, rather than relying on production of all the Earth return propellant on the surface.

NASA's *Human Exploration of Mars Design Reference Architecture 5, Addendum 2*, lists key tenets influencing mission planning (Table 9.1). Reading through this list shows what a lasting influence Mars Direct has had on mission architecture.

Table 9.1 Key tenets in Mars mission design

TENET	EXPLANATION	COMMENT
First human mission by 2030	Fundamental assumption behind design reference.	National policy projected for mid 2030s
Pre-deployment of cargo	Sending cargo ahead reduces mission mass and risks; cargo should be sent at least one launch window (opposition) ahead of the main mission, potentially in 2028.	Agency decision
Long-stay (conjunction) missions	Provide the best balance between performance, mission return, cost and risk. They have the lowest mission masses and also the smallest level of variation between mission opportunities.	Agency decision
Minimise exposure to deep space	Radiation and zero-g exposure to be reduced. Long-stay missions allow transit durations $\sim$180 days, while maximising the time on the surface. The low gravity (0.3 g) on Mars may help ameliorate for zero-g transits.	Major issue to be resolved
In situ resource utilisation	Generating propellant and oxygen from Mars' atmosphere provides significant architectural leverage. It reduces total mission mass and the mass of the lander. Abort to orbit during landing is not possible as Mars has an atmosphere, hence landing without ascent propellant is not an issue.	Agency decision
Heavy lift launches	Results from DRA 5 indicate that surface Mars missions require between 7 and 11 launches at 110–140 tonnes per launch. The exact number and mass depending on the propulsion systems selected for the transfer flight.	Major issue related to other mission decisions
Payload to Mars surface $\sim$40 tonnes	Providing the ability to transport 40 tonnes to the surface gives the optimum balance between lander size and the number of landers needed.	
Advanced propulsion	Advances in propulsion technology are required, including extended storage of cryogenic propellants. Nuclear thermal propulsion and aerocapture could reduce total mission mass and number of launches. The extent to which advances are required is dependent on the speed of transit to Mars required.	Important factor in reducing total mission time; DST projected to use combination of chemical and electrical propulsion

Overall, NASA's plans for the coming decades evolved from the Constellation programme and the budget overruns and delays that dogged that initiative. The combination of Orion, SLS, Gateway and DST provides a flexible system where the components support each other and mitigate compromises made at various stages. Politics has certainly been a factor in this and critics cite congressional insistence on using past space shuttle suppliers and the lack of clear scientific justification for Gateway as limiting factors. However this plays out, private enterprise and independent groups such as Zubrin's Mars Society continue to study ways to get to Mars and survive on the surface and may well provide competition for NASA. An important player in the private sector is SpaceX.

9.5.3 SpaceX

It will start off building just the most elementary infrastructure, just a base to create some propellant, a power station, blast domes in which to grow crops—all of the sort of fundamentals without which you cannot survive … . And then really there's going to be an explosion of entrepreneurial opportunity because Mars will need everything from iron foundries to pizza joints. I think Mars should really have great bars: the Mars Bar.

Elon Musk[18]

Elon Musk founded Space Exploration Technologies Corp. (SpaceX) in 2002, with the specific aim of reducing launch costs and enabling the colonisation of Mars, in order to ensure the long-term survival of humanity.

Since its founding, the company has developed the Falcon range of launch vehicles, culminating with Falcon Heavy (Section 9.3.1), along with the *Dragon* family of reusable spacecraft. Dragon 1 vehicles have ferried supplies to the ISS since 2012 and in 2017 a Dragon launched that reused a substantial number of components that had flown on a previous mission. Dragon 2 is capable of carrying a crew. It had its first successful automated flight to the ISS in March 2019 and a manned flight is anticipated later in the year.

SpaceX have set a target launch price for crewed flights of $160 million, which works out at $20 million per seat, if the ship carries its maximum crew of seven. The figures are conditional on NASA ordering four flights per year and compares with the $76 million per seat Russia charges for Soyuz flights. In October 2014, Dragon 2 was selected, along with Boeing's *CST-100 Starliner*, for development as part of NASA's programme to identify commercial opportunities for delivering crews to the ISS. SpaceX was awarded $2.6 billion to support development; Boeing received $4.2 billion.

In a significant 2017 announcement, SpaceX revealed that Japanese billionaire Yisaku Maezawa purchased tickets for a 2023 flight around the Moon on its Starship. Maezawa plans to invite six to eight artists from various disciplines to join his week-long journey. The funds raised for the trip will help finance further development of the vehicle and the Super Heavy first stage intended to support Mars missions.

The Super Heavy rocket will be 58 m high and develop 53,000 kN of thrust. It will be completely reusable, manoeuvring around to a powered landing as a complete stage.

The 48 m long Starship (Figure 9.12) will be equipped with a combination of vacuum and sea level engines, to allow it to return to Earth under power and be reused. Unlike the space shuttle, Starship will land vertically in an effort to reduce the time needed for turnarounds.

The engine specifications and landing technique are key aspects of the design as the craft is eventually intended to land on Mars, where the atmosphere is far too thin to support a glide flight path as used by the space shuttle. If Starship can land on Earth using rocket motors, the far lower gravity of Mars should not present a problem. The Martian atmosphere can be used to remove the bulk of the energy on entry before moving to the powered phase of the descent.

Musk has even considered using Super Heavy with a version of Starship to provide intercontinental flights on Earth by employing sub-orbital hops just outside the atmosphere. This could reduce the time for the 9,648 km journey from London to Hong Kong from 12 hours to 34 minutes. He believes that rocket reliability can rival that of commercial airliners.

The payload area of Starship can be set up to deliver satellites and crews to orbit or crews and payloads to the Moon, or even Mars. The configuration for a Mars flight will feature a pressurised volume of 825 m^3

Figure 9.12 An artist's impression of SpaceX's Starship separating from the Super Heavy first stage. (Image courtesy of Space Exploration Technologies Corp.)

to accommodate 40 cabins, a galley, storage and a radiation shelter for protection from solar storms. Musk claims that Starship can support a team of 100 on a Mars mission.[19]

Payload capacity to LEO is expected to be 140 tonnes. If refueling in orbit becomes a reality, the payload capacity can be increased further. Four launches of Starships configured as tankers would allow an already in-orbit craft to be refuelled to capacity, enabling 140 tonnes of payload beyond LEO with a ΔV ~6 km/s. Alternatively, using NASA's projected 40 tonnes of payload to Mars from the design tenets, a refuelled Starship would have a ΔV ~8 km/s available which would allow a transit time to Mars of ~180 days on a long-stay mission profile (Figure 4.15).

Starship's reusability and increased payload capacity results from the use of advanced materials in its construction, for example, a cryogenically formed stainless steel alloy. The dry mass of the craft will be 85,000 kg.

The current design for a heat shield[20] calls for a stainless steel 'sandwich' on the 'windward' side consisting of two layers of alloy cross-braced with stringers and with a liquid (either water or fuel) circulating through the internal volume. The outer layer will be perforated with tiny holes through which the liquid will bleed out. Hence, cooling of the heat shield will occur by a transpiration process. The craft should enter Mars atmosphere at 7.5 km/s and decelerate aerodynamically. Its heat shield is designed to cope with multiple re-entries, but at this speed the engineers expect some ablation, similar to brake pad wear on a car.

Flexibility is a crucial feature of the Super Heavy–Starship system. In order to release resources for the new design, SpaceX will need to use early Super Heavy–Starship vehicles as replacements for their current launch fleet. The company intends to build a stock of Falcon 9s and Dragons to enable space missions to continue during a phased transition to Super Heavy–Starship, which will then take over commercial satellite deployment, space station supply and crew deliveries.

In January 2019, SpaceX completed the external skin of its first test vehicle (*Starhopper*) and moved to developing the internal plumbing of the fuel tanks and supply lines. A successful test firing with a single engine took place in April 2019 while the craft was tethered to the ground. A series of test flights are planned for altitudes from 500 to 5,000 m to verify liftoff and vertical landing capability. SpaceX is also in the process of constructing the first orbital prototype.

Musk's plans for Mars have become clearer with his recent statements along with those of Paul Wooster, the company's principle Mars development engineer. The aspirational (although Musk believes realistic) target is to land two automated cargo Starships on Mars in 2022 (Figure 4.20, this is an opposition, but on the upward side of the distance cycle). These missions would confirm the suitability of the landing site based on factors such as the availability of water (which is frozen under the surface on Mars). Some of the base infrastructure, such as landing pads, would be robotically constructed.

The next step is launching two more cargo ships and two crew ships in 2024. The task of the first mission is to develop the best source of water and the second will build a propellant plant powered by a solar array (Mars Direct's influence is seen here as well). With six ships on the surface, the makings of a sustainable base are in place.

As each ship will carry 'at least 100 tons worth of supplies', crews should be able to stay on Mars for years. In the initial phases, the ships will likely serve as habitats. The amount of supplies transported should also provide equipment redundancy. SpaceX is admittedly focussed on getting to Mars as soon as possible and trying to avoid the development costs and time required for new technologies. As these come 'on stream' they will be incorporated into future missions as the base expands. Note that the plan requires vehicle refueling, but does not call for any assembly in orbit.

At some stage, it will be necessary for the ships to return to Earth to enable SpaceX to recoup some costs. Eventually Musk wants to offer one-way tickets to Mars for $200,000. He envisions a permanent 'new frontier' settlement. Much of SpaceX's development and planning has focussed on getting to Mars, but their first Mars Workshop in 2018 was attended by 60 scientists (including some from NASA). Musk is seeking expertise and possibly commitments from other individuals and organisations to develop or provide means to keep crews healthy during long-duration flights and while they establish settlements on the surface of Mars.

Some of Musk's more far-reaching speculations read like science fiction, but with an extensive canon of material on Mars colonisation, based on reasonable scientific principles, to draw on, this does not mean that he is engaged in flights of fancy. His company's implicit philosophy that it is okay to fail during development has served it well up to now.

Perhaps in this day and age, the spirit of President Kennedy's original Apollo dream is more easily found in wealthy entrepreneurs than it is in government-funded bureaucracies.

ENDNOTES

1. Design Reference Architecture 5.0, 2009.
2. Excerpt of Michael Griffin's speech to the American Astronautical Society, November 2005.
3. https://www.gao.gov/new.items/d09844.pdf
4. http://www.bbc.co.uk/worldservice/news/2010/02/100201_obama_moon_wup_sl.shtml
5. https://www.nasa.gov/feature/nasa-seeks-us-partners-to-develop-reusable-systems-to-land-astronauts-on-moon
6. https://www.nasa.gov/topics/moon-to-mars/overview
7. This commercial rocket is constructed by the United Launch Alliance, a joint venture of Lockheed and Boeing.
8. Message to the SLS and Orion workforce from J Bridenstine, NASA Administrator, March 14, 2019. https://blogs.nasa.gov/bridenstine/2019/03/14/a-message-to-the-workforce-on-sls-and-orion/
9. DU-1310 screens from Honeywell.
10. https://spacenews.com/nasa-considering-flying-orion-on-commercial-launch-vehicles/
11. This is known as the restricted three-body problem, the idea being that the third body is too small (in mass) to disturb the motions of the other two. Their mutual orbit is a solution of the two-body problem and the third body moves within that context.
12. This discussion should really be couched in terms of the centre of gravity of the Mass 1–Mass 2 system, as both objects orbit that location.
13. That is, 90 degrees to the plane of the Earth–Moon system.
14. https://www.nationalreview.com/2017/05/nasa-lunar-orbit-space-station-terrible-idea/
15. Future logistics missions will probably fill the number gaps.
16. *Free Press,* updated edition, June 28, 2011. ISBN-10: 145160811X.
17. Lava tubes on Mars were discovered in orbital images.
18. https://www.businessinsider.fr/us/spacex-meeting-mars-mission-planning-workshop-2018-8
19. https://www.liebertpub.com/doi/10.1089/space.2018.29013.emu
20. https://www.popularmechanics.com/space/rockets/a25953663/elon-musk-spacex-bfr-stainless-steel/

Appendix 1

GLOSSARY

Apocynthion	Point on an orbit about the Moon where the distance to the centre of the Moon is the greatest.
Aphelion	Point on an orbit about the Sun where the distance to the centre of the Sun is the greatest.
Apogee	Point on an orbit about the Earth where the distance to the centre of the Earth is the greatest.
Astronomical unit (AU)	Unit of distance used by astronomers that corresponds to the mean distance between the Earth and the Sun: 1.496×10^{11} m.
Avionics	Combination of aviation and electronics.
Booster	General term for first stage of a multiple stage rocket; carries no payload or certain components found in solid rocket boosters on the space shuttle.
Burn	Firing of spacecraft or rocket engine.
Cape Canaveral	Military missile facility and test next to the Kennedy Space Center on the Florida coast.
Cape Kennedy	No such place. *See* Cape Canaveral or Kennedy Space Center.
Composite	Combination of two or more materials, e.g. fibreglass is a combination of glass fibres and a resin bond. Composites are created to use specific properties of their component materials to produce a single unit better suited to a task than the individual materials would be.
Conjunction	Any two celestial bodies are in conjunction when a third reference body lies on an extension of the line joining the two bodies. The term most commonly refers to the position of the Sun and one of the outer planets as observed from Earth. See also *opposition*.
Cryogenic	Ultra-low temperature. Cryogenic liquids must be stored at extremely low temperatures or they will boil into gases. Liquid hydrogen and liquid oxygen are commonly cryogenic liquids used in space flight. They must be stored at $-250°C$ and $-180°C$ respectively.
Electrolyte	Liquid through which electricity can be conducted, commonly the liquid components of wet cell batteries and fuel cells. Electrolytes conduct electricity as they contain ions in solution.
EVA	Extra Vehicular Activity; activity that takes place outside a spacecraft in space. *EVA suit* is another term for a space suit.
Fuel	Component of propellant used to power a chemical rocket motor. The fuel reacts chemically with an oxidiser to produce a high temperature exhaust.

Golf	A game based on directing a small white pellet into a small round hole from a great distance with implements designed to make the task as difficult as possible.
Ion	An atom that gains an extra electron and becomes negatively charged or an atom that loses an electron and becomes positively charged.
Kennedy Space Center	NASA launch facility on the Florida coast from which the Apollo flights were launched and the space shuttle took off and landed.
Laser	Device producing a concentrated beam of light that is *coherent* (all waves in the beam are in phase) and nearly *parallel* (the beam diverges only due to diffraction). *Laser* is an acronym for Light Amplification by Stimulated Emission of Radiation.
Mach number	Ratio between the speed of an object moving through a fluid and the speed of sound in the fluid. A Mach number greater than 1 indicates that an object is moving at hypersonic velocity. The speed of sound in air at sea level and 15°C is 340 m/s
Newton (N)	SI unit of force. Engine thrust is often quoted in pounds. To convert pounds into N, calculate the correct number of kilograms and multiply the result by 9.8 N/kg (the strength of gravity on the Earth's surface) to yield force. A one-pound force equals 4.4 N.
Opposition	Any two celestial bodies are in opposition with respect to a third when they lie on diametrically opposite sides of the third body. See *conjunction*.
Oxidiser	Component of propellant used to power chemical rocket motors. The oxidiser is needed to react with the fuel in a chemical reaction similar to the way substances burn in the atmosphere.
Pascal (Pa)	SI unit of pressure corresponding to a 1 N force over an area of 1 m². Normal atmospheric pressure at sea level is ~100 kPa (or ~about 15 psi).
Pericynthion	Point on an orbit about the Moon where the distance to the centre of the Moon is the smallest.
Perigee	Point on an orbit about the Earth where the distance to the centre of the Earth is the smallest.
Perihelion	Point on an orbit about the Sun where the distance to the centre of the Sun is the smallest.
Plasma	Very high temperature gas in which each atom or molecule has been electrically charged by loss of electrons. At low temperatures the electrons will again bind with the atoms or molecules.
Propellant	Combination of fuel and oxidiser used to propel a chemical rocket. Mixing the two components generates a chemical reaction yielding an exhaust gas characterised by high temperature and pressure. In more advanced engine designs, the propellant is a single liquid (e.g. liquid hydrogen) that is heated to boil and escape from the engine or ionised and ejected as a stream of charged particles.
Sintering	Process by which a material containing a fibre or powder is heated below the melting point of the fibre or powder and compressed to fuse the fibres together.
Specific impulse	Quantity used to rate propellant performance. Specific impulse (measured in seconds) is the thrust produced per unit weight of propellant consumed per second. This is equivalent to the time needed to burn 1 kg of propellant to produce 1 N of thrust.
Stack	Connected stages of a rocket or modules of a spacecraft.
Telemetry	Continuous stream of data relayed to ground by a spacecraft covering parameters such as system performance, course and thrust.
Thrust	Rocket engineering term for force produced by an engine. Thrust can be calculated by multiplying the speed at which the exhaust leaves the engine by the rate at which propellant is consumed. The result is measured in Newtons.

Appendix 2

DERIVING FORMULAS

> I have just worked out various aspects of the problem of ascending into space with the aid of a reaction machine, rather like a rocket … The scientifically verified mathematical conclusions indicate the feasibility of an ascent into space with the aid of such machines, and, perhaps, the establishment of settlements beyond the confines of the Earth's atmosphere.

Konstantin Tsiolkovsky

Konstantin Tsiolkovsky (1857–1935) was a mathematics teacher in the small Russian town of Kaluga. Inspired by the books of Jules Verne, he set out precise mathematical calculations to demonstrate some of the basic (and now accepted) principles of rocketry:

- Liquid fuel rockets present the best potential for exploring space and the planets because they can be throttled, stopped, and restarted. Rockets powered by solid propellants cannot be used in such a flexible manner.
- Liquid fuel rockets are also more efficient because the hotter and lighter the exhaust gases, the more efficient the engine.

Tsiolkovsky suggested that liquid hydrogen could be burned to produce hotter and lighter exhaust gases. He was also the first person to consider concepts similar to specific impulse, and the first to calculate the escape velocity from Earth. He also devised a way to stack rocket components to use thrust more efficiently and called it the principle of the *sky train*. We now employ this principle in staged rockets (such as the Saturn V) and discard strap-on boosters when no longer needed (as in the space shuttle).

Chapter 3 discussed staging as a more efficient way of lifting payloads into orbit based on the rocket equation:

$$\Delta V = uLn\left(1 + \frac{M_p}{M_R}\right)$$

but did not show the proof of the equation. We now turn to this derivation.

The starting point for studying the motion of any system in which the mass changes must be Newton's second law of motion in the form:

$$F = \frac{d}{dt}(\text{momentum of system})$$

Written in Chapter 3 as:

$$F = \frac{d}{dt}(mv) = m\frac{dv}{dt} + v\frac{dm}{dt}$$

This is not a very useful form because it assumes that every part of the system (represented by total mass m) suffers the same change of velocity. This is not the case for a rocket. The fuel mass ejected as exhaust undergoes a change in velocity different from that of the rocket and remaining fuel. A far better approach is to consider the momentum change of the elements of the total system.

Consider a system composed of a rocket and fuel to be ejected. If M is the mass of the rocket (including the stored fuel) and Δm the mass of fuel to be burnt at a time interval Δt, the change in momentum of the rocket is:

$$(M)(v + \Delta v) - Mv = Mv + M\Delta v - Mv = M\Delta v$$

where Δv is the amount by which the velocity of the rocket changes over time (Δt).

The momentum of the burnt fuel also changes. Initially this element of fuel moved forward with the rocket at velocity v. At the end of the time interval, it moves forward with velocity $(v - u)$ where u is the velocity of the fuel relative to the rocket. If this seems strange, remember that our view is from the ground. If the rocket moves fast enough, the exhaust exiting backward will be moving forward relative to the ground.

Final momentum $-$ initial momentum $= \Delta m(v - u) - \Delta mv = -\Delta mu$

$$\therefore \text{Rate at which the fuel momentum is changing} = -u\frac{\Delta m}{\Delta t}$$

which as $\Delta t \to 0$ becomes $-u\dfrac{dm}{dt}$.

$$\therefore \text{Total rate of change of momentum of the system (rocket + fuel element)} = M\frac{dv}{dt} - u\frac{dm}{dt}$$

So:

$$F = M\frac{dv}{dt} - u\frac{dm}{dt}$$

where F is the external force applied to the system. In a rocket moving through deep space far from any sources of gravitational pull, the net force acting on the system is zero.

$$0 = M\frac{dv}{dt} - u\frac{dm}{dt}$$

$$\therefore M\frac{dv}{dt} = u\frac{dm}{dt}$$

The only way that this equation can be solved is to relate M and m. Fortunately, this is simple as dm/dt represents the rate at which the burnt fuel mass *increases*, which must be the same as the rate at which the rocket's mass is *decreasing*.

$$\therefore \frac{dM}{dt} = -\frac{dm}{dt} \to M\frac{dv}{dt} = -u\frac{dM}{dt}$$

Rearranging and integrating:

$$-\int_{M_R + M_P}^{M_R} \frac{dM}{M} = \frac{1}{u}\int_{v}^{v + \Delta v} dv$$

and hence:

$$-[LnM]_{M_R + M_P}^{M_R} = \frac{1}{u}(v + \Delta v - v)$$

$$\therefore uLn\left(\frac{M_R + M_P}{M_R}\right) = \Delta v$$

or:

$$\Delta v = uLn\left(1 + \frac{M_P}{M_R}\right)$$

The situation is quite different if the rocket is in a region where there is a gravitational force. Then the rate of momentum change equation becomes:

$$F = M\frac{dv}{dt} - u\frac{dm}{dt} = -Mg$$

$$\therefore \frac{dv}{dt} - \frac{u}{M}\frac{dm}{dt} = -g$$

Rearranging gives:

$$\frac{dv}{dt} + \frac{u}{M}\frac{dM}{dt} = -g$$

$$\therefore \int_{v}^{v+\Delta v} dv + u\int_{M_P + M_R}^{M_R}\frac{dM}{M} - -g\int_{0}^{t} dt$$

$$\Delta v + uLn\left(\frac{M_R}{M_R + M_P}\right) = -gt$$

Hence:

$$\Delta v = -uLn\left(\frac{M_R}{M_R + M_P}\right) - gt = uLn\left(\frac{M_R + M_P}{M_R}\right) - gt = uLn\left(1 + \frac{M_P}{M_R}\right) - gt$$

TWO-STAGE EFFICIENCY

As the text box explaining multiple stage rockets in Section 3.6 indicates, the ΔVs from the first and second stages are:

$$\Delta V_1 = uLn\left(\frac{N}{n(1-R) + RN}\right)$$

$$\Delta V_2 = uLn\left(\frac{n}{R(n-1)+1}\right)$$

After the second stage burn is complete, the total ΔV is:

$$\Delta V = \Delta V_1 + \Delta V_2 = uLn\left(\frac{N}{n(1-R) + RN}\right) + uLn\left(\frac{n}{R(n-1)+1}\right)$$

To maximise this for different values of n, the mathematical steps require differentiating the expression with respect to n and setting the result equal to zero.

$$\frac{d(\Delta V)}{dn} = \frac{-u(1-R)}{n(1-R)+RN} + \frac{u}{n} - \frac{uR}{R(n-1)+1} = 0$$

Combining these fractions together over a common denominator is tedious:

$$\frac{-u(1-R)n[R(n-1)+1] + u[n(1-R)+RN][R(n-1)+1] - uRn[n(1-R)+RN]}{n[n(1-R)+RN][R(n-1)+1]} = 0$$

For this expression to equal zero, the numerator must be zero, so we will focus our algebraic massaging on that.

Provided $u \neq 0$, we can cancel that factor out of every term and start to expand the rest:

$$(nR - n)[Rn - R + 1] + (n - Rn + RN)[Rn - R + 1] - Rn(n - Rn + RN) = 0$$

Continuing the expansion and cancelling terms:

$$(R^2 n^2 - R^2 n + Rn - Rn^2 + Rn - n)$$
$$+ (Rn^2 - Rn + n - R^2 n^2 + R^2 n - Rn + R^2 Nn - R^2 N + RN)$$
$$- (Rn^2 - R^2 n^2 + R^2 Nn) = 0$$

Giving:

$$RN - R^2 N - Rn^2 + R^2 n^2 = 0$$

Factorising:

$$R^2(n^2 - N) - R(n^2 - N) = R(R-1)(n^2 - N) = 0$$

So that either, $R = 0$, $R = 1$ or $n = \sqrt{N}$ – the last option being the physically relevant solution in this case.

Index

Note: Page numbers followed by "*n*" with numbers indicate endnotes.